# HITLER'S
# FORGOTTEN
# CHILDREN

# HITLER'S FORGOTTEN CHILDREN

## THE SHOCKING TRUE STORY
## OF THE NAZI KIDNAPPING CONSPIRACY

### INGRID VON OELHAFEN
### & TIM TATE

WITH DR DOROTHEE SCHMITZ-KÖSTER

First published 2015 by
Elliott and Thompson Limited
27 John Street
London WC1N 2BX
www.eandtbooks.com

This paperback edition first published in 2017

ISBN: 978-1-78396-318-8

9 8 7 6 5 4 3 2 1

A catalogue record for this book is available from the British Library.

Jacket design: Jem Butcher Design
Typesetting: Marie Doherty
Printed CPI Group (UK) Ltd, Croydon, CR0 4YY

*This book is dedicated to all the victims of
Nazi Germany – men, women and, above all,
children – and to those throughout the world today
who suffer from the persisting evil which teaches that
one race, creed or colour is superior to another.*

# CONTENTS

# PREFACE

Blood runs through this story. The blood of young men spilled on the battlefields of war; the blood of civilians that ran through the gutters of cities, towns and villages across Europe; the blood of millions destroyed in the pogroms and death camps of the Holocaust.

But blood, too, as an idea: the Nazi belief – absurd as this seems today – in 'good blood', precious Ichor to be sought out, preserved and expanded. And with it, the inevitable counterpart: 'bad blood', to be ruthlessly eradicated.

I was born in 1941 in the depths of the Second World War. I grew up in its wake, and under the shadow of its brutal progeny, the Cold War.

My history is the history of millions of ordinary German men and women like me. We are the victims of Hitler's obsession with blood, as well as the beneficiaries of the post-war economic miracle that transformed our devastated and pariah nation into the powerhouse of modern Europe. Our story is that of a generation raised in the shadow of infamy, but which found a way to struggle towards honesty and decency.

But my own story is also that of a much more secret past, still cloaked in silence and shrouded in shame.

I am a child of Lebensborn.

Lebensborn is an ancient German word meaning 'fountain of life', twisted and distorted by the word-smelters of National Socialism. What did it mean in the madness of Nazism? What does it mean today? My search for the answers – to uncover my own story – has taken me on a long and painful road: a physical journey that has led me across the map of modern Europe. It has been an historical expedition, too: an often uncomfortable return to the Germany of more than seventy years ago, and into the troubled stories of those countries overrun by Hitler's armies.

The journey has also forced me to make a psychological voyage into everything I have known and grown up with: a fundamental questioning of who I am, and what it means to be German. I will not pretend that this is a simple story: it will not always be easy to read. But neither has it been easy to live.

I am not, by nature, overtly emotional. The expression of emotion, such a commonplace thing in twenty-first-century society, does not come easily to me. I have spent my life attempting to suppress my inner self, to subordinate my feelings to the circumstances in which I have grown up, as well as to the needs of others.

But this is a story which, I believe, needs to be heard. More, much more, it needs to be understood. It is not unique, in that there are others who have endured much of what has shaped my life. But while I share a common thread with thousands of others who passed through the vile experiment of Lebensborn, to the best of my knowledge no one else shares the particular twists of fate, history and geography that have defined my seventy-four years on earth.

Lebensborn. The word runs through my life like the blood coursing through my body. To see it, to understand it, demands much more than a superficial examination. The search for the roots of this story requires a deep and intrusive investigation of the most hidden places.

We must start in a town and a country that no longer exist.

# ONE | AUGUST 1942

*'Men ... must be shot, the women locked up and
transported to concentration camps, and the children
must be torn from their motherland and instead
accommodated in the territories of the old Reich.'*
REICHSFÜHRER-SS HEINRICH HIMMLER, 25 JUNE 1942

**Cilli, German-occupied Yugoslavia, 3–7 August 1942**
The schoolyard was crowded. Hundreds of women – young
and old – clutched the hands of their children and found what
space they could in the packed courtyard. Nearby, Wehrmacht
soldiers, rifles slung over their shoulders, looked on as the families slowly drifted in from towns and villages across the area.

These women had been summoned by their new German
masters, ordered to bring their children to the school for 'medical tests'. Upon arrival they were arrested and told to wait.
Otto Lurker, commander of the police and security services
for the region, watched relaxed and impassive – his hands
resting comfortably in his pockets – as the yard filled with families. Once, Lurker had been Hitler's gaoler: now he was the
Führer's leading henchman in Lower Styria. He held the rank
of SS-Standartenführer – the paramilitary equivalent of a full
colonel in the army – but that summer's morning he was casu-
ally dressed in a two-piece civilian suit.

Yugoslavia had been under Nazi rule for sixteen months. In March 1941, with the surrounding countries of Hungary, Romania and Bulgaria having recently joined the Reich's alliance of Balkan nations, Hitler put pressure on the kingdom's ruler, the Regent Prince Paul, to fall into line. He and his cabinet bowed to the inevitable, formally tying Yugoslavia to the axis powers, but the Serb-dominated army launched a coup d'état, replacing Paul with his seventeen-year-old second cousin, Prince Peter.

News of the revolt reached Berlin on 27 March. Hitler took the coup as a personal insult and issued Directive 25, formally designating the country an enemy of the Reich. The Führer ordered his armies 'to destroy Yugoslavia militarily and as a state'. A week later, the Luftwaffe began a devastating bombing campaign while divisions of Wehrmacht infantry and tanks of the Panzer Corps swept through towns and villages. The Royal Yugoslavian Army was no match for Germany's Blitzkrieg troops: on 17 April the country surrendered.

The occupying troops immediately set about fulfilling Hitler's instruction to dismantle all vestiges of the state. Some 65,000 people – primarily intellectuals and nationalists – were exiled, imprisoned or murdered, their homes and property handed over to their new German masters. The Slovene language was prohibited.

But for the rest of 1941 and throughout the first half of 1942, partisan groups, led by the communist Josip Broz Tito, fought a determined campaign of resistance. Germany retaliated with a brutal crackdown: the Gestapo swooped on fighters and civilians alike, deporting thousands to concentration camps across the Reich. Others were executed as a warning against resistance. In the nine months following September 1941, 374 men and women were lined up against the walls of the prison yard at Cilli and summarily shot. Photographers recorded the murders for the purposes of both posterity and propaganda.

On 25 June 1942, Heinrich Himmler – the second most powerful and feared man in Nazi Germany – issued orders to his secret police and SS officers for the elimination of partisan resistance.

> This campaign possesses every required element to make harmless the population which has supported the bandits and provided them with human resources, weapons and shelter. Men from such families, and often even their relatives, must be shot, the women locked up and transported to concentration camps, and the children must be torn from their motherland and instead accommodated in the territories of the old Reich. I expect to be provided with a special report on the number of children and their racial values.

Against this bloody backdrop, 1,262 people – many the surviving relatives of those executed as an example to the rest of the population – assembled in the schoolyard that August morning to await their fate.

Among them was a family from the nearby village of Sauerbrunn. Johann Matko came from a family of known partisans: his brother, Ignaz, had been one of those lined up and shot against the wall of Cilli prison in July. Johann had been dragged off to Mauthausen concentration camp. After seven months in the camp he was allowed to return home to his wife, Helena, and their three children: eight-year-old Tanja, her brother Ludvig – then six – and nine-month-old baby Erika.

When all the families were accounted for, an order was

given to separate them into three groups – one each for the children, the women, and the men. Under Lurker's direction the soldiers moved in and pulled children from the grasp of their mothers; a local photographer, Josip Pelikan, recorded the harrowing scene for the Reich's obsessive archivists. His rolls of film captured the fear and alarm of women and children alike: his shots included scores of toddlers held in low pens of straw inside the school buildings.

As the mothers waited outside, Nazi officials began a cursory examination of the children. Working with charts and clip-boards, they painstakingly noted each child's facial and physical characteristics.

These, though, were not 'medical tests' as any doctor would know them: instead they were crude assessments of 'racial value' which assigned each youngster to one of four categories. Those who met Himmler's strict criteria for what a child of true German blood should look like were placed in Category 1 or 2: this formally registered them as potentially useful additions to the Reich population. By contrast, any hint or trace of Slavic features – and certainly any sign of 'Jewish heritage'– consigned a child to the lowest racial status of Categories 3 and 4. Thus branded as *Untermensch*, their value was no more than future slave labour for the Nazi state.

By the following day this rudimentary sifting had finished. Those children deemed racially worthless were handed back to their families. But 430 other youngsters, from young babies to twelve-year-old boys and girls, were taken away by their cap-tors. Marshalled by nurses from the German Red Cross, they were packed into trains and transported across the Yugoslavian border to an *Umsiedlungslager* – or transit camp – at Frohnleiten, near the Austrian town of Graz.

They did not stay long in this holding centre. By September 1942, a further selection had been made – this time by trained 'race assessors' from one of the myriad organisations established by Himmler to preserve and strengthen the pool of 'good blood'.

Noses were measured and compared to the official ideal length and shape; lips, teeth, hips and genitals were likewise prodded, poked and photographed to sort the genetically precious human wheat from the less-valuable chaff. This finer, more rigorous sieving re-assigned the captives within the four racial categories.

Older children newly listed in Categories 3 or 4 were shipped off to re-education camps across Bavaria in the heartland of Nazi Germany. The best of the younger ones in the top two categories would – in time – be handed over to a secretive project run by the Reichsführer himself. Its name was Lebensborn and among the infants assigned to its care was nine-month-old Erika Matko.

TWO | **YEAR ZERO**

*'It is our will that this state shall endure
for a thousand years. We are happy to
know that the future is ours entirely!'*
ADOLF HITLER: *TRIUMPH OF THE WILL*, 1935

At 2.40 a.m. on Monday, 7 May 1945, in a small red-brick schoolhouse in the French city of Reims, Generaloberst Alfred Jodl, Chief of the German Armed Forces High Command, signed the unconditional surrender of the Thousand Year Reich. The five terse paragraphs of this act of capitulation handed over Germany and all its inhabitants to the mercy of the four victorious Allied powers – Britain, America, France and Russia – from 11.01 p.m. the following night.

A week earlier, Hitler and most of his inner circle had committed suicide in the bowels of the Berlin Führerbunker. Heinrich Himmler – Hitler's chief henchman and the man in charge of the entire Nazi apparatus of terror – was on the run, disguised in the coarse grey serge of an enlisted soldier and equipped with forged papers proclaiming him to be a humble sergeant.

It was over: six years of 'total war' in which my country had murdered and plundered its way across Europe. Now we had to live with the peace.

Who were we then, on that May morning? What was Germany – once the begetter of Bach and Beethoven, Goethe and Schiller – in the aftermath of the brutality of the Blitzkrieg, let alone the Final Solution? What would peace look like to the victors and to the vanquished?

A new term was coined to describe our situation in 1945: *Die Stunde Null*. Literally translated, this means 'zero hour' but for the smouldering remains of Germany – a country of ruins, shame and starvation – it was more accurately 'Year Zero': both an end and a beginning.

What did it mean to be a German from 11.01 p.m. on Tuesday, 8 May 1945? To the Allies – the new owners of every metre of turf and of every individual life from the Maas in the west to the Memel in the east – it meant subjugation, suspicion and suppression. Never again, said the four occupying powers, would the poisonous twin rivers of German nationalism and militarism be allowed to rise up and flood the continent. Within hours there would be mechanisms and procedures in place to enforce this ideal – systems that, though I was too young to know then, would direct the course of my life.

To Germans, this question of identity meant something different. Something much less philosophical, something that could perhaps be categorised as the three Ps: physical, political and psychological. Of these, the greatest – the most pressing – was undoubtedly the physical.

Germany in May 1945 was a wasteland of blown-up bridges, damaged roads, burned-out tanks. In the dying weeks and months of his Reich, consumed by madness and impotent rage, Hitler had issued orders to create 'fortress cities'. The Fatherland was to be defended to the last drop of pure German blood and the last brick of German building. There was to be no surrender but, instead, a *Götterdämmerung* of flame and sacrifice to mark the final days of his self-proclaimed Master Race.

The result was less a noble funeral pyre than a thousand-mile-wide bonfire of his vanity. Forced to fight for every inch of territory – and bludgeoned by Allied carpet-bombing – Germany was reduced to a post-apocalyptic desert. Piles of rubble lay where buildings had once stood: in Berlin alone there were seventy-five million tons of it piled up along and across almost every street. Other German cities suffered equally, obliterated by bombing and house-to-house fighting that damaged or left derelict seventy per cent of their buildings. And everywhere, now hollow and haggard, a once-proud people who had subjugated those they believed to be inferior.

Newsreels and photos (Allied ones, since the German press had been shut down from the moment of surrender) captured previously unimaginable scenes. Clustered around half-destroyed buildings, blown apart so that the remnants of a once normal life were exposed for all to see – a fireplace, shreds of wallpaper, the remains of a toilet – were the living ghosts of women and children. Orphans, refugees, the aged and the wounded: everywhere a dystopian tableau of anonymous bodies lying dead in the street, watched – or more often avoided – by skeletal figures who might well soon join them.

All of Germany, at least in the cities, was picking through debris, creating makeshift shelters, scrounging for food and either hiding from or fearfully fraternising with the victorious occupying armies. Not from choice, but from necessity.

In the last weeks of the war, the country's economy – so long directed by and for the benefit of the Nazi Party – had collapsed as badly as its buildings. Ironically, there was plenty of money, but coins and paper bills were useless: as every available resource was diverted away from the people to the needs of the army, and as explosions ripped up the railway network, preventing what food was harvested from being distributed, there was little or nothing to buy with the now-useless marks.

Nor did Germany's new masters appear to have a coherent idea of what to do with it. Between July and August 1945, the Allied leaders – Churchill (and, later, Attlee), Truman and Stalin – met at Potsdam to plan the future. Unlike the end of the First World War, when Germany was defeated and subjected to severe punishment and reparations but not wiped from the geographical and political map, the decision was taken that the country would cease to exist once the war ended. In its place would be four separate 'Occupation Zones', each owned and ruled by one of the war's victors, according to its own principles and plans.

Yet beyond that there had been little concerted thinking about what, practically, would be done with the former German state once Hitler had been defeated. France had favoured breaking the Reich into a series of small independent states while America had considered returning Germany to a pre-industrialised nation focused and dependent on farming. Washington would come to relent, to accept that requiring tens of millions of Germans to live as medieval peasants was unworkable as well as undesirable. But the Allies failed to contemplate how their separate occupations would function, or to address the monumental problem of feeding both a conquered people – a population swelled by more than ten million refugees from the east – and the massive armies imposing the peace.

There was simply not enough food – and without a functioning transport system, what little there was couldn't be moved to the places where it was most needed. Worse, there was a widespread feeling among the occupying armies that the Germans were long overdue a taste of their own medicine: had the Nazi rampage across Europe not deliberately starved villages, cities, entire nations to the point of death?

This, then, was Hitler's true legacy: a nation starving to death; a population reduced to a desperate struggle for survival, subsisting at best on half the calories needed to sustain life. A country not simply beaten and half-destroyed but wiped completely out of existence.

I was three and a half when peace came. A small, quiet and archetypally blonde German child, I lived in Bandekow, a tiny hamlet in the rural heart of the Mecklenburg region, with my mother, grandmother and slightly younger brother Dietmar. Our home was a big farmhouse, half-timbered and characteristic of the region, set in acres of forest. We were, I think, typical both of a particular class of pre-war Germans and, by contrast, of the post-war country at large. On both sides our family was old, well established and, notwithstanding the wrecked economy, well off.

My mother, Gisela, was the daughter of a shipping line magnate from Hamburg. The Andersens belonged to the old Hanseatic class – the patrician and prestigious ruling elite which had made its money and its name from trade since Hamburg was declared a free city by the 1815 Congress of Vienna.

Our house in Bandekow had been in my mother's family for generations: it belonged to my great uncle, but had almost certainly been used as a country retreat in the years before 1945. Certainly, the Andersens kept their main residence in Hamburg itself and my grandfather remained there, with my grandmother dividing her time between the two homes.

Gisela was one of four Andersen children. Her brother had been killed, serving in the Wehrmacht in the last days of the war; her eldest sister was estranged – the result of some unspoken act of dishonesty that tarnished the otherwise respectable family name – but her remaining sibling, my Aunt Ingrid (known universally as Erika, or 'Eka'), was a constant companion in my childhood. At the end of the war, Gisela was thirty-one. She was young, bright – in the brittle and privileged way of her class – and pretty. She was also married, though not, as it turned out, happily.

Hermann von Oelhafen was a career soldier. He had served with honour in the First World War: he was seriously injured in 1914, again in 1915, and, after a final wound in 1917, was awarded the Iron Cross for his pains. Like Gisela, he came from an aristocratic background: both his father and mother could boast the tell-tale 'von' – the mark of the upper class – in their family names.

But where Gisela was young and lively, Hermann was the complete opposite. He was thirty years older than Gisela and suffered from severe epileptic seizures. Whether these were the cause of his peevish, mean-spirited nature I do not know: what I am certain of is that their marriage – which took place in 1935, during the first confident years of Hitler's reign – was, by 1945, effectively over. As I grew from a toddler to a young child, I rarely saw my father: we lived in the farmhouse at Bandekow, while Hermann lived 1,000 kilometres away in the Bavarian town of Ansbach.

Perhaps outwardly there was nothing very strange in a married woman living alone with her children and mother. In this our family was typical of the now-dissolved German nation in the immediate months after the war: most adult men – even the very young and the elderly – had been drafted into military service and were now either dead, missing or held in prisoner of war camps across Europe. Germany was a country – more accurately, a former country – of women and children.

But though it played its part, the war was not the prime reason for the separation of my parents. There was an unbridgeable gulf between them; an emotional fracture even less tractable or open to resolution than the divisions imposed upon their nation. I was too young to know it at the time, but it would render my childhood as bleak as the deteriorating political situation in which we found ourselves.

Politics. The second 'P' which defined life at the end of the war. Not politics as modern generations have come to know and disregard it; not the jockeying for position and power between rival parties in a settled democracy: politics in 1945 was truly red in tooth and claw.

The last days of the war had seen the Allied forces smashing their way through Germany from all points of the compass. American tanks and troops rolled eastward from France, Belgium and Holland; the British fought their way northwards, up through the country from Italy and Austria; and the vast armies of the Soviet Union raced westwards from what had, before the war, been Poland. For each there was an overriding imperative to conquer and control as much German territory as possible: whatever they held when the war finally ended would, under the Potsdam Agreement, become their property with little prospect of subsequent redistribution. In those last weeks of spring 1945, the borders of post-war Europe were being claimed and, at the same time, the seeds of the Cold War were being planted.

When the fighting was over, it turned out that my father's home was in the American zone: henceforth his fate would depend on the way Washington saw its duties and rights over the territory it now owned. Bandekow, however, was in the Soviet occupation zone, and Moscow had very different ideas about how to dismantle the infrastructure of Nazi Germany – as well as what it wanted to do with its share of the former Reich.

Initially, at least, there was agreement between the Allies on the need to bring Hitler's surviving henchmen to justice. A four-power war crimes tribunal was established to put the National Socialist machine on trial; Göring, Jodl, Hess, von Ribbentrop and twenty other leaders of the National Socialist state were locked up in cells beneath the Palace of Justice in Nuremberg to await trial for crimes of war and crimes against humanity. Other than Hitler and Goebbels, the most notable absentee from this roll call of infamy was Himmler, creator of the SS and mastermind of the Nazi's apparatus of terror: after being captured he had committed suicide before he could be transported to Nuremberg.

The eventual trial and conviction of almost all these men was undoubtedly a triumph for justice, but it also marked the high point of cooperation between the occupying powers. After Nuremberg, America, Britain, France and the Soviet Union would each take a radically different approach to the land and populations they controlled: the individual fates of tens of millions of former Germans depended on which zone they happened to have been in when the war ended. Very soon these great political divides would change the lives of our little family for ever.

The contrast between the four occupying powers was played out first in the way they viewed Nazi Party members. Denazification was a phrase coined in Washington during the last years of war: President Franklin Roosevelt and his successor, Harry Truman, recognised that the party's tendrils had wound themselves throughout every aspect of German life, from the political to the judicial, the public to the personal. In May 1945 there were more than eight million members of the Nazi Party – around 10 per cent of the total population. What was to be done about this entwining of the mechanics of fascism with the warp and weft of everyday life?

The search for an answer was not confined to America, of course. Each Allied power faced the problem of how to pull out the roots of National Socialism while ensuring that its own zone of occupation kept functioning. The first step was to outlaw the party. On 20 September 1945, Control Council Proclamation No. 2 announced that 'The National Socialist German Workers Party (NSDAP) is completely and finally abolished and declared to be illegal' throughout the former Reich.

But the party itself was only the most visible of a byzantine tangle of Nazi organisations. Beneath it were more than sixty other official associations, ranging from internationally notorious bodies like the SS, Gestapo and Hitler Youth to more obscure societies (even within Germany) such as the Reich Committee for the Protection of German Blood and the Deutsche Frauenschaft, the National Socialist Women's Movement. All were duly made illegal: more importantly, previous association with any one of them would be enough to mark someone as a possible Nazi sympathiser.

Neither Hermann nor Gisela were – to the best of my knowledge – Nazi Party members. I never heard them express fascist opinions or support for Hitler. But their personal histories (my father as a career soldier, who had been a desk officer in the Wehrmacht for much of the war; my mother as a former member of Deutsche Frauenschaft) must have led to some investigation by the denazification officials of their respective Occupation Zones.

The Americans were initially fiercely committed to denazification, but quickly became the most pragmatic of the occupying armies. Washington's military government realised that, however desirable, widespread purges of suspected Nazis would mean that the entire responsibility for organising day-to-day life fell exclusively on its shoulders — a burden that, for a war-weary nation anxious to bring its troops home, was simply too onerous.

And so while my father, like every adult living in the American zone, was required to fill out a questionnaire (termed variously a *Fragebogen* or a *Meldebogen*) in which he affirmed that he had never been a member of any Nazi organisation, there was little follow-up or detailed examination of these self-declarations. With little or no oversight, most applicants were issued with official documents pronouncing them to be 'good Germans', free of the stain of fascism. They quickly became known as *Persilschein* — pieces of paper that were able to wash the past as clean as any soap powder.

The Soviet approach was very different. Perhaps because it had suffered greater losses and devastation than any of the four Allied powers — or, more likely, because Stalin had clear plans for the future of the Soviet zone — Moscow adopted a much less relaxed approach.

The Soviet Military Administration in Germany — known by its acronym, SMAD — controlled a vast swathe of territory from the Oder river in the east to the Elbe in the west. On April 18, 1945, Lavrenti Beria, Stalin's much-feared head of secret police, issued order Number 00315: it mandated the immediate internment of active Nazis and senior members of Party organisations. No investigations were required prior to these arrests. Ultimately, 123,000 Germans were rounded up and incarcerated in ten special camps set up across the Soviet zone.

The existence of these prisons – run by the NKVD, Stalin's equivalent of the Gestapo, and frequently on the site of former Nazi concentration camps – was in itself a secret. No contact was allowed between prisoners and the outside world, but inevitably word did leak out: the often random nature of arrests and internment (by February 1946, genuine Nazi Party members formed less than half of the total number of prisoners), and fear of being dragged off to the network of *Schweigelager* (literally, 'Silence Camps') weighed heavily on an already fearful German population under Soviet military rule.

Almost anything – anonymous denunciation, previous membership of an obscure Nazi society or contact with anyone in the other three Occupation Zones – was enough to earn a knock on the door and transport to a *Schweigelager*. All too often this proved to be a one-way ticket: almost 43,000 men and women would die behind the barbed wire of these post-war concentration camps.

Did my mother worry about the risk that her involvement with Deutsche Frauenschaft posed to our household in Bandekow? I do not know: the von Oelhafens were a close-lipped family, rarely given to discussion of emotions, much less those of the past. It would be many years before I discovered the secret at the heart of my childhood, a secret that tied Gisela, Hermann and me to a sinister Nazi organisation, one which would certainly have spelled trouble for us if SMAD came to hear of it.

Was this an added worry, clouding my mother's mind? Again, I do not know. What I do know is that as summer turned to winter, Gisela was terrified of something else: rape.

Throughout 1945, as the Soviet Army fought its way into Germany, its troops mastered one phrase above all: *Komm, Frau*. It was an order that brooked no disobedience and led to the same inevitable conclusion. Tens of thousands – perhaps ten times that number – of German women paid, with their bodies, the price for Hitler's brutal treatment of Russian cities and populations. Rape was so commonplace in the Soviet sector that the question for many women, of all ages, was not *whether* they had been violated but *how many* times.

It was also quasi-officially sanctioned. Although SMAD commanders in some parts of the Occupied Zone paid lip service to stamping out the violation of German women, in reality others paid a heavy price for doing so. One young Red Army captain, Lev Kopelev, intervened to stop the gang rape of a group of girls and was sentenced for his troubles to ten years in a labour camp: a tribunal convicted him of the crime of 'bourgeois humanism'.

It was, of course, true that neither the internment camps nor rape were confined to the Soviet sector. The Americans imprisoned thousands of suspected Nazis, often in appalling conditions for years, and French troops frequently ravaged German women in cities under their control. But in the final months of the war, Hitler and Goebbels had fanned the flames of national fear by issuing a constant stream of propaganda about the brutality of the Red Army – and from the moment they fought their way onto German soil, the Soviet occupiers fulfilled the worst of these predictions.

Our family was as vulnerable as any, if not more so. My mother and my Aunt Eka were young and pretty and we came from the hated bourgeoisie: our home was large, comfortable and well-stocked with food from the farm, but it was also isolated, and my brother was the only man in the household. The fear of rape hung over us as the winter wore on. My mother would later remember – one of only a sparse handful of personal feelings she ever shared with me – hiding under the bed whenever she heard rumours that Red Army soldiers were in the area.

But however debilitating the fear, in truth we were better off than most of the population in the Soviet Zone. We had a roof over our heads, unlike the vast majority of people in the bombed-out cities. The winter of 1946–7 was one of the harshest in living memory: temperatures plummeted to -30° and for the millions struggling to exist in the bombed-out basements of their former homes there was no protection from the biting cold. And since what remained of the rail network after the final, disastrous months of fighting was rapidly dismantled by the Soviet Army and taken back east as war reparation, there was little coal to be had: thousands of people simply froze to death.

But it was food – or rather, the lack of it – that soon became the overriding preoccupation. German ration cards were no longer valid: whatever limited provisions had previously been available were now being claimed by SMAD to feed the Red Army. In cities across the country, hunger joined fear as the measure of existence.

In the areas under Moscow's control, new rationing measures were introduced. The Russians created a new five-tier system: the highest level was reserved, bizarrely, for intellectuals and artists; the next level down was assigned to the women – *Trümmerfrauen,* as they were called – who worked in chain gangs, tearing down and clearing semi-derelict buildings, often with nothing more than their bare hands. This was much more valuable than the official wages of 12 Reichsmarks they received for cleaning up every thousand bricks. Hard physical labour was the only way to survive and, in the ruins of the nation, Germany's women dug for the salvation of their families.

The levels of rationing below this fell incrementally and dramatically. The lowest card, nicknamed the *Friedhofskarte* (literally meaning 'cemetery ticket'), was issued to those who performed no useful function in the eyes of our Soviet masters: housewives who did no work and the elderly.

Two new words joined the lexicon of post-war lives that winter. The first was *Fringsen*: it emerged after the Catholic cardinal of Cologne, Josef Frings, gave formal blessing to what many of his flock were already doing – stealing in order to survive. Crime rose dramatically: in addition to the uncountable tally of thefts and rapes by Red Army soldiers, Germans under Soviet occupation began preying on each other. Berlin alone averaged 240 robberies and five murders every day. Urban crime may not have been a pressing concern for the von Oelhafens, living in the relative security of rural Mecklenburg, but the second new word had a very real meaning. *Hamstern* meant, quite literally, 'to hamster': in practice, it was a constant procession of city dwellers to and from the countryside, desperate to trade their few remaining possessions for the food we had in relative abundance.

This was the reality of *Stunde Null*: an existence defined by three constant companions: fear – especially of the Red Army and of its determination to exact revenge on German civilians for Hitler's war – hunger and cold. This was Germany, my country and my life on my fourth birthday. This was the legacy of the glorious Reich. And there was worse in store. Throughout 1946, as relations between the occupying powers worsened, Moscow's intentions towards those under its rule in

SMAD grew starker. As well as stripping the zone of wealth and food, it began the process of removing the one, flickering hope we had enjoyed when the war ended: freedom.

The boundaries between the four zones were becoming ever less passable. An 'Inner German Border', as SMAD termed it, had been established around the Soviet-held territory in July 1945, but since then it had been only sporadically policed. Although anyone wanting to move between the Soviet sector and the other Allied Occupation Zones officially needed an *Interzonenpass,* at least one and a half million Germans had managed to flee into the American or British zones. Now that began to change.

In the summer of 1947, preparations were underway for the eventual transformation of SMAD into the new communist-ruled German Democratic Republic. New contingents of Soviet soldiers were assigned to the official border checkpoints. Unofficial crossing places would soon be blocked by newly dug ditches and barbed wire barricades. The Cold War was beginning and we were living on the wrong side of the coming Iron Curtain. In the summer of 1947 my parents – separated both physically and emotionally – made a remarkable joint decision. It was time to escape.

*'Ingrid is very brave and overcomes
the strenuous walk without complaining.'*
GISELA VON OELHAFEN'S DIARY, JUNE 1947

My mother kept a journal. Unknown to me – she never told me about it, even when I was an adult – she jotted down the barest of details of my early years across a sparse handful of pages. This slim, black leather-bound notebook contains all I know about my first eight years of life.

It begins with a small black and white photo of me, three years old, barefoot and wearing shorts, captioned: 'Bandekow – Ingrid, Summer 1944'. Over the page is an envelope dated June 4, 1944 and containing – according to my mother's note – a few strands of my hair. If that seems fairly conventional, the sort of journal any loving mother might keep as a record of her daughter's childhood, the rest of the content fails to match that impression. There are very few entries – no more than four or five for each of the five years my mother wrote in it. And the nature of the inscriptions themselves are curious: they are all in the third person. Gisela refers to herself as '*Mutti*' (the German colloquial word for Mummy) throughout, never 'I'.

This sort of third-person journal was, I have since gathered, not uncommon in Germany during the later war years – perhaps she did it to make it easier for her children to read later. But given that she never told us about it, that benign explanation seems unlikely. Her curiously detached writing style seems rather to emphasise the difficulty my mother found in conforming to a maternal stereotype, and the distance I always felt between us.

Still, the journal does give me some idea of what sort of child I was. The entry for my birthday on 11 November 1944 reads:

> Today Ingrid is three years old. At her age she is not tall, but she thrives and prospers wonderfully and has a healthy constitution. She has a strong will and a disposition to violent temper. Her character is calm and persistent. She does not pay attention to people she does not know: at that point her little ego takes centre stage and claims a little too much of her small world.

The next item, dated a month afterwards, hints at a desire to win my mother's affections. For reasons undisclosed, I was left alone with Dietmar at lunchtime; when my mother returned she noted that:

> Ingrid – with a serious face – was very busy feeding her brother just as Mummy always does.

If the diary is any guide, I appear not to have succeeded. Throughout the twelve months of 1945, my mother only managed to put pen to paper on five occasions: two to record the effect of measles on me, one noting the happy news that I was no longer afraid of our family dog, and two more in which my slowness to speak ('She doesn't make complete sentences, her maximum is three or four words') is observed and preserved. There was, as my mother must have known perfectly well, a very good reason why I might be slow to get my childish tongue around German words. But there is no mention of this in the pages of the notebook.

Nor does she dwell on what must have been a traumatic period in my life the following year. In the summer, my mother laconically recorded that I (and presumably Dietmar) had been sent away to a children's home, more than 250 kilometres away at Lobetal, near Berlin. How did we get there? I do not know: her diary is silent on this, as on so much else. All that she wrote was that:

> Mummy is ill – meanwhile Ingrid lives from 5 August to
> 1 November in a children's home in Lobetal. There she
> suffers from mumps – but not too badly.

My mother's illness, I learned decades later, was in fact a nervous breakdown. Perhaps it was the collapse of her marriage and the burden of looking after two small children. Perhaps it was the strain of living under Soviet occupation; the constant fear of arrest or – worse – rape by the Red Army. The first entry in the notebook for 1947 shows that she had made up her mind to escape – and that she had enlisted my father, though they were still estranged, into her dangerous plans.

> 1 May 1947. Daddy takes both children to the children's
> home in Lobetal. Mummy wants to cross the border
> illegally.

I cannot pretend that I was ever close to my mother, nor can I claim to have ever really felt from her the sort of love that a child should take for granted from a parent. Gisela also plainly knew this; another terse diary entry in my mother's handwriting noted that I was always much fonder of my grandmother. 'Granny is loved over all others, often more than Mummy. She gets on with the children very well.' But even so, I have to acknowledge that the decision to make a bid for freedom was immensely brave.

The border between what would, less than two years later, become the German Democratic Republic and the British zone of occupied Germany was both political and physical. It was, of course, forbidden to leave the Soviet zone without a special permit, and these were far from easy to obtain. Even writing the idea of crossing illegally in her diary could – had it been discovered – have led to interrogation, imprisonment in the Silence Camps, or worse.

In addition to this, the journey to the border was as arduous and complicated as it was dangerous. Bandekow might have been less than fifteen kilometres as the crow flies from the Elbe river, which marked out much of the boundary with the British zone, but there was no way to cross it. My mother had already made a secret trial run and must have discovered that the nearest bridges at Lauenburg and Dömitz had been blown up by the retreating German army in 1945. The nearest bridge left intact was 150 kilometres further south at Magdeburg.

With the country's railways still in chaos and with private cars (let alone the petrol to run them) a rarity, getting to Magdeburg would have been a challenge for a healthy adult, travelling alone. My mother was far from well – and she would have two very young children to drag with her every step of the way: it must have been a daunting prospect. Encumbered by Dietmar and me, she could not carry anything with her: the three of us would make the trek in whatever clothes we had and – if successful – would arrive in the safety of British territory with nothing more than the clothes we stood up in.

The near-impossibility of getting simply from one place to another in 1947 is clear from the convoluted escape route my mother carefully wrote down in the diary. Tracing it now on a map, the first part of the journey took us east, not west: ever deeper into the Soviet zone and away from the sanctuary we sought. We began on 30 June, travelling – I think – by horse and cart, twenty-five kilometres to the little town of Lübtheen. Here, my mother found a hotel to put us up for the night and in which we could await the arrival of her co-conspirator the next morning.

I have no idea how my father managed to get papers to cross from the American zone into Soviet-controlled territory, nor how he obtained the car into which the four of us crowded that morning. All I know is that the thirty-kilometre journey, eastwards to the city of Ludwigslust, was the last time our family was together.

The reason for heading so far east was waiting for us in the station at Ludwigslust. Both the platform and the train which would take us back westwards to Magdeburg must have been very crowded. That summer more than ten million refugees and released prisoners of war were on the move: like us, many were desperately trying to find a route out of the Soviet zone. Somehow – I was never told how – we had precious train tickets: my mother's diary records only that the train was so crowded my father had to push the two of us into her arms through the train window. She makes no mention of the fact that he did not make the journey with us, staying behind on the platform to (so I like to imagine) wave farewell to his wife and children.

Magdeburg was more than 150 kilometres south, and the train journey took all day. When we finally arrived, it was evening and we must have been both tired and hungry. Finding food could not have been an easy task: Magdeburg had been heavily bombed in 1945 and by the time we arrived it was still a city of ruins and ruination. And although it was in the Soviet zone, our Soviet-issued ration cards were not valid there. Alone with two small children in a strange and devastated city, my mother took the only available option: she found a black-market trader and handed over sixty marks for a few pieces of bread.

There is no information in my mother's diary about where we stayed that night. In the chaos that was Magdeburg, it seems unlikely that we would have found a hotel: it says simply that we stayed in the city all the next day, changing accommodation in the evening to be closer to the next stage of our path to freedom.

We had, first, to get a train out of Magdeburg, heading north to the village of Gehrendorf. Here the little river Aller was the boundary between east and west. On the other side was the hamlet of Bahrdorf, safely inside the British zone. All that lay between us and sanctuary were the slow-running waters of the Aller. But there were no boats and no bridges: the only way across was to wade. So that was what we did.

It cannot have taken too long, for the Aller is small at the best of times, and in the height of summer would have been reasonably shallow. Nonetheless it must have been challenging for a fraught young woman with two young children in the heat of a midsummer's day. She must have been scared, hoping not to be seen by Red Army border guards and praying that neither Dietmar nor I would cry out and give our position away. The only record I have is what my mother later set down in her notebook:

> The temperature is very high. Ingrid is very brave and overcomes the strenuous walk without complaining.

Finally we reached the sanctuary of the other side. We crawled up the bank and after a long trek through No Man's Land, we reached the British zone. We were free.

My mother could not have known it – though, given the urgency and determination inherent in her succinct account of our flight from east to west, she must have sensed the Iron Curtain beginning to fall – but we had made our getaway just in time. By September 1947, the borders between the Soviet Zone and those of its former western allies were closely guarded by a new influx of NKVD troops; it was not long before orders were given to shoot would-be escapers on sight.

What did freedom look like to Gisela von Oelhafen that summer's evening? What did it mean to her to have reached safety after two years under Soviet occupation, and to have escaped with her children from Moscow's iron rule? I wish I could ask her now.

༄

A day's hard travelling later, we arrived in Wunstorf, a little town just west of Hanover and the penultimate stop on my mother's journey to her family home in Hamburg. I say my mother's journey quite deliberately because she would make the last leg of her trek alone. Her diary entry – terse as ever – recorded the very different fate allocated to Dietmar and me: '4 July: To Loccum, children's home.'

She had taken us out of the Soviet zone and into the less dangerous territory of the British sector. But that is as far as her maternal protection extended. No sooner had she spirited us away to safety, than she sent us away. My second night of freedom ended in the surroundings of a home for unwanted children.

I would spend the next six years, lonely and isolated, in the care of the church. In fact, my new life began exactly as the old life had ended: cold and frightened.

# FOUR

# HOME

*'Dear Mummy, please take me home for ever.
I'm longing for you and Granny and Aunt Eka.'*

LETTER TO MY MOTHER
FROM THE CHILDREN'S HOME

My first real memory is an orange. I have snatches of other, possibly earlier, recollections – lying, cold, under a blanket on the floor of a train; a line of camp beds in a long room and a rat running over my feet – but the first actual memory I know to be true is of the orange. I am at a long wooden dining table in a big room. There are a lot of people, grown-ups and children. I know that many of the adults are homeless men and women who have been invited here for the day; the children, though, live in this building. Each of us is given a plate with fruit on it, including a single orange as a special treat.

I know where and when this memory comes from. The year was 1947 and I was almost six years old. The room with the long table was in the children's home to which Dietmar and I were dispatched. It was Christmas Day.

The home was run by the Protestant church and was called Nothelfer, which, literally translated, means 'help us in affliction'. There were sixty-five boys and girls living there, all under the age of ten. Some were displaced persons – children who had lost their parents during the war or in the chaotic mass migrations of the immediate post-war months. Dietmar and I were different: we had two living parents who knew where we were but who, for reasons best known to themselves, had sent us to be cared for by others.

We were physically as well as psychologically isolated. Nothelfer was on Langeoog, a small island in the North Sea ten kilometres from the coast of mainland Germany and 200 kilometres from Hamburg. To be fair to my parents, I don't think they had intended to send us so far away: when we first arrived in July, it had been situated near Hanover. But at some point in the subsequent five months those premises were closed and we were moved up to Langeoog.

Given its location, it was hardly surprising that Nothelfer was cold. I can still feel the wind whipping up sand from the island's long beach, seemingly stripping the skin from my legs and arms.

The home was staffed by sisters from the religious order and at times the regime could be harsh: physical punishment was part of our daily routine. If we were disobedient, if we wet the bed, if we broke the cardinal rule forbidding us to slide down the sand dunes, we were spanked. One by one we had to line up and pull our pants down and one of the sisters would beat our bare bottoms with a stick.

We would stay here for four years. From time to time our parents made the journey out to the island to visit us. Their visits were rare and they never came together, always alone. My father had moved from the American Occupied Zone into the British sector, and was building himself a new house in the Westphalian spa resort of Bad Salzuflen. Although he and my mother were separated, they had not divorced (and would never do so). Occasionally they spent a little time together – generally when Dietmar and I were allowed out to visit one of them – but my mother had begun to make a new life for herself in Hamburg.

Immediately after sending us to Nothelfer, she had returned to her family's home in the city, a large three-storey building in an exclusive neighbourhood. Number 39 Blumenstrasse had three floors and a basement, and large gardens that ran down to Rondeelteich – one of the big lakes in the heart of Hamburg where people went boating, sailing or swimming.

There she lived with her mother, my Aunt Eka, and a housekeeper-cum-cook. Initially, they also shared the house with British Army officers who had been billeted there at the end of the war. Their presence was, ostensibly, the reason why Dietmar and I had been bundled off to the children's home: according to my mother there was not enough room for us all.

Whether or not this was actually true (since the position didn't change when the soldiers moved out), it did enable her to start a new life, unencumbered by either husband or two small children. She enrolled at college and began training to become a physiotherapist: once qualified, she turned one of the ground-floor rooms into a surgery where she worked with a growing clientele of patients.

She also took advantage of her essentially single status to find a boyfriend. Neither Dietmar nor I would ever meet this man, but within two years she gave birth to a baby boy, whom she named Hubertus. He was not, I am certain, my father's child, but he was formally registered as a von Oelhafen.

I cannot, in all honesty, say that my father's visits to Langeoog meant a great deal to us. Whether this was because of his age or his stiff and disciplinarian nature I do not know: what I do recall with awful clarity is the heartache of being apart from my mother, and the loss I felt after her occasional trips to see us. I missed her terribly.

Among her effects, years later, I found a note from one of the kinder sisters who ran the home.

Highly honored madam,

I'd like to add a few sentences to the letter Ingrid has written to you. For the past few weeks I have been worrying about Ingrid. She yearns intensely for her 'Mummy'. Every day she talks about 'Mummy' or asks: do you think I'm allowed once to go to Mummy? I'd like so much to see her. Aunt Emi, do you think that I would be allowed to leave the island and stay for a short time with Mummy?

Ingrid is eating very little and feels miserable. In my opinion, the reason for this calamity is her longing for Mummy.

In school Ingrid is one of the best. She is hard-working. In general she is a nice child.

I feel committed to inform you about this.

Best wishes from Schwester [Sister] Emi

Did she ever reply to this letter, or to any of the letters I wrote to her? I don't recall: and yet she kept at least some of them. Along with Sister Emi's letter, I found this undated note, scrawled in my childish hand after an all-too-brief visit to see her.

Dear Mummy,

Thanks a lot for the parcel. I am only writing a little today.
Dear Mummy, please take me home for ever. I'm longing
for you and Grandmother and Aunt Eka. I always cry if
someone talks about you, or if I think of you. On the trip
[back to Langeoog] I couldn't eat. I still have the chocolate
and the two Deutschmarks.

Please write as fast as possible to the authorities and say
that I'm allowed to leave the island soon. Christa [another
girl in the home] will leave the island this month or next.
But I want to leave this month!

Dietmar told me he has got a lot of fruits and sweets.
Dietmar teases me a lot and asks: why didn't you stay
in Hamburg? I want so much to leave the island. Dear
Mummy, please arrange that I can leave. Christa told me,
she had cried too when she had to go away from her
parents.

Many greetings and kisses from Ingrid. Don't write
to Daddy about me having written to you. Dear, dear
Mummy, please come and pick me up.

She never did. And even the sporadic, terse entries in the little
journal she had – apparently – kept for me ceased abruptly in
the summer of 1949.

I was ten years old when I finally left Langeoog for good in
1952. I had passed the exams to go to *Mittelschule*, the interme-
diary level of education between elementary and senior schools,
and had high hopes that at last I would be allowed to live with
my mother in Hamburg. Instead my father sent for my brother
and me: we were to share his new home in Bad Salzuflen.

By this point Hermann von Oelhafen was sixty-eight years old, bitter at the loss of his wife, suffering from poor health and utterly ill-equipped to look after young children he barely knew. Of the ten years of my life and nine years of Dietmar's, he had lived with us for a matter of months at most. It was surely a little too late to start being a father.

I believe I know now why Hermann ordered us to Bad Salzuflen. I think my father still loved Gisela and he hoped our presence there might somehow draw her back to him; that despite her evident love affair with another man – and their child – Dietmar and I would be the glue that mended their broken marriage.

In this, as in much else, he was to be disappointed. My mother came occasionally to visit us (a spare bedroom in the smart, if far from grand, house on Akazienstrasse was maintained for her exclusive use), but there was never any question of reconciliation.

From the moment we arrived, life in Bad Salzuflen was horrible. Even as a young child, Dietmar was spirited and difficult, though not particularly naughty. Today I think he might have been diagnosed with ADHD: certainly he and Hermann fought a battle of wills. He was routinely late coming home from school – even though he never seemed to want to learn there – and Hermann, who was hot-tempered at the best of times, had no understanding or patience for this irritating little boy. Very soon he began to beat Dietmar.

It was frightening to watch: on one occasion he physically threw him across the room. And yet somehow Dietmar wasn't frightened of him. I, by contrast, was terrified: even though my father never hit me, I lived in fear of his temper. I began to rely on Dietmar to ask for Hermann's permission to do even the smallest things. One day we wanted to go swimming but I didn't dare ask. Dietmar went straight away in my place, and permission was granted, but it didn't change anything for me. I was still too scared to speak to my father. And then Dietmar was taken away.

Someone – presumably the children's welfare department of the local government – decided that since Hermann did not live with his wife and that there was therefore no mother in our house, Dietmar could no longer live with us.

There were a number of odd aspects to this decision. For a start it only applied to Dietmar: although I was less than a year older than him, the authorities did not appear to believe I had to be removed from my father's care. Officially the reason for this discrepancy was that I was to be looked after by the middle-aged couple who lived with us as Hermann's cook-cum-housekeeper and general help (the Hartes). But no one explained why the paid-for care of Emmi and Karl Harte was good enough for me yet insufficient for Dietmar.

More puzzling still was the revelation that Dietmar had family – a completely different family from ours – living in Munich. Had I been older I would, of course, have realised that the nine months between his birthday and mine meant that we were unlikely to have had the same mother. But even if I'd understood that, I would never have anticipated the truth. It was a complete shock to discover – aged ten – that the boy I had always known as my brother had in fact been fostered by my parents.

And so it transpired that Dietmar had an uncle, an aunt and a sister – blood relatives all – who had, presumably, been looking for him. I don't remember anyone explaining how he had originally come to live with us instead of them: one day he was simply taken away from Bad Salzuflen. As it turned out, he never rejoined his long-lost family: instead – for reasons I have never understood – they placed him in the care of another children's home.

I missed Dietmar very badly. And I was now on my own (save for the Hartes) in my father's house and, if my letters to Gisela are anything to go by, absolutely terrified.

22th June 1952

Dear Mummy!

Please send me some envelopes and stamps. Dear Mummy, please pick me up this week and bring me to Hamburg, I'm not able to stay longer with Daddy. I must tell you that my fear of him is greater than before. He told me off once because I've cried about you. Now I cry every day.

Dear Mummy, please pick me up immediately, I cannot stand it here with Daddy any longer. Or come and stay here for ever. But 'Uncle Harte' says you are afraid of Daddy too. Dear Mummy, perhaps you can send the stamps to 'Uncle Harte', but don't tell Daddy that I've written to you.

Mummy, we can organise it this way: You come and pick me up forever and explain to Daddy you had written to Munich [to the child welfare department] to ask whether they would allow me to stay with you. You [explain that you] didn't bring the letter with you, but promise to send it when we get back to Hamburg.

And in Hamburg we'll write a letter on typewriter and send it to Daddy and pretend it has come from Munich. I'd like so much to come to you just this week, please pick me up quickly. Today I cried again because I have been thinking of you. I'm not in the mood to play because you are not here. Please pick me up at the 25th of June. Greetings and kisses from Ingrid. Please pick me up the 25th. Please, please dear Mummy.

These pleas went unheeded. Although my mother continued to come for short visits, she never took me home with her. Whether this was because my father forbade it or because she didn't want me to live with her, the outcome was the same: I was effectively imprisoned in the house in Bad Salzuflen with an increasingly bitter and parsimonious old man.

Even as an eleven-year-old, I was aware that my father was less well off than my mother. He received a state pension from the new West German government; reward for his years serving both the Kaiser and the Reich as an officer in the army. But this did not, for example, seem to be sufficient to pay the price of a daily newspaper to be delivered. Instead he would make the journey into town where he would stand outside the newspaper offices and read the morning's edition, which was always pasted up in the window. Occasionally I was permitted to go with him.

Hermann's health was deteriorating. He had suffered from epilepsy for many years (a condition he had apparently concealed from Gisela when asking her to marry him). Now it became increasingly severe. Although I never saw him have a full-blown grand mal fit, when a seizure took him he became 'absent' – completely lost within himself. There was no way to communicate with him, and his behaviour was strange and frightening. Often he picked up a knife and waved it around wildly. Once he was hospitalised and while he was away, Frau

Harte was more generous than he was with the marmalade for my breakfast. When Hermann came home again he saw how much had been used from the jar and became angry. I was punished for my evident greed by being denied marmalade for a week.

School became my refuge. I had made friends with other children and was fortunate that their parents, perhaps seeing how unhappy I was at home, were kind and loving towards me. I loved spending time with them, seeing in their lives the thing that I valued – and missed – most: a real family. And then, when I was eleven, I discovered that I was not who I thought I was.

I woke up one morning and found that I couldn't open my eyes. My father took me to the doctor's surgery.

We sat in the reception area, waiting for my turn. When the doctor called out the name 'Erika Matko', my father stood up and led me into the consulting room. He handed over my health insurance card and I saw that it too had the name 'Erika Matko' printed on it.

I had no idea why I was being called by a different name. But I didn't dare say anything to the doctor or my father; I was still too frightened of him to question anything. At the end of the consultation I was prescribed a course of sun lamps – a common enough treatment in those days for vitamin deficiency (most likely a problem dating back to my years in the children's home at Langeoog) – and we went back home. Nothing was said about that different name, but I had not forgotten.

Shortly after that I had a conversation with Frau Harte. Every Friday it was our routine to clean the house together, and I could talk freely to her about whatever I had on my mind. It was the closest thing I had to a normal relationship with an adult. As we were polishing, I asked her if she knew why my name was written down as Erika Matko.

Emmi told me that Hermann and Gisela were not my biological parents. She said that when I was a baby they had fostered me, just like Dietmar, and that my original name was Erika Matko. Emmi wasn't embarrassed to tell me that I had been fostered. The war had fractured so many families and left so many children without parents that our situation was far from unusual.

I don't recall being upset at discovering the truth about myself. I was not close to Hermann and I think that I processed the information by deciding that it explained his coldness towards me, and why I was not allowed to live with Gisela.

But of course I wondered where I had come from. I assumed that my real parents were German – it never occurred to me to think otherwise – and I speculated about what had happened to them. Perhaps they had been in prison; maybe they had died in the war. Emmi said she had wondered whether I was originally Jewish because of my prominent nose. But although my father had told her that I was a foster child, she didn't know any more than that. Everything else was just speculation.

Of course I never said anything to Hermann. Nor, the next time Gisela came to visit, did I ask her about it. But Hermann must have told her about the visit to the doctor's surgery and I assume she felt she had to say something. She started to tell me that I was fostered and how she had fetched me from a children's home but I quickly cut her off, saying 'I know.' I don't really know why I stopped her: perhaps it was my childish way of showing her that it was all too late, that I was hurt that she had kept the truth from me. The subject was never mentioned again.

The one person I would have liked to talk to was Dietmar. We had been close in the children's home and in the few months we had spent together in Hermann's house. But by then he had been taken away and I had no way to contact him: I didn't even have an address that I could write to.

Life carried on as before. Every morning I went to school – where I was registered and known as Ingrid von Oelhafen – and returned in the afternoon to the house in Bad Salzuflen and the man I now knew was not my biological father, of whom I was still very afraid.

∞

Over the next two years, Hermann's heath continued to deteriorate and he was often still in bed when I left for school. I would go into his room to wish him good morning, but in truth this was no more than living up to my duties as a good daughter.

Then, one morning in April 1954, towards the end of the spring term and with the long summer holidays approaching, I said goodbye to him as usual. I noticed that he seemed a little disorientated when I left, but I didn't say anything to the Hartes because I assumed it was just another symptom of his illness. When I came back from school he was in a very bad way: it was clear he had had a stroke. My father – or rather my foster father, as I now knew him to be – was taken to hospital and died two weeks later.

I have to admit that I was not sad. I felt happy to be free of him and his harsh, unforgiving ways. And I assumed that at long last I would be allowed to live with Gisela in Hamburg. What did hurt me was Emmi and Karl's reaction: they criticised me severely for not telling them about Hermann's condition that morning.

My high hopes for a new life with my mother – I still thought of her as 'Mummy' then, even though I knew I wasn't her 'real' child – were not to be: or not immediately, at least. Gisela was too busy with her thriving physiotherapy practice and her five-year-old son, Hubertus.

For six long months I carried on living in Hermann's home with the Hartes looking after me. It was not until nearly October 1954 that I was finally sent to Hamburg. And by then, the strange story of Erika Matko and my true identity seemed to have been forgotten.

*'The lost identity of individual children is the social
problem of the day on the continent of Europe.'*
INTERNATIONAL REFUGEE ORGANISATION
INTERNAL MEMORANDUM, MAY 1949

When I was fifteen years old I saw my face on a poster in the street. A decade after the end of the war, and seven years after the formation of our new Federal Republic, Germany was still a nation of displaced and unclaimed children. United Nations agencies had spent those years searching across Europe for close to two million missing boys and girls, separated from their parents by bombings, military service, evacuation, deportation, forced labour, ethnic cleansing, or murder. By 1956, it had traced just 343,000 of them.

The Red Cross had decided that one way to discover the origins of children who may have been brought to Germany during the war was to post photos of the children as they were then in newspaper advertising columns. Underneath these lists of faces and names ran the headline: *'Who knows our parents and our origins?'* They also pasted up large posters on columns and lamp posts on streets across West Germany. It was from one of these, in the centre of Hamburg, that my younger face peered back at me.

It was, to say the least, a shock. I had no idea that anyone was looking for me, nor how they would have obtained my photograph. I had to presume that Gisela had given it to the authorities, but no one had said anything about it to me.

By that point I had been living in my mother's house on Blumenstrasse in Hamburg for two years. Two years during which my dreams of a happy family life had proved to be no more than an unrealistic and childish fantasy. I had spent half of my young life longing to be with my mother, aching to feel loved and looked after. By the time I saw my photograph on the poster, reality had set in – and set me in my ways.

I knew, of course, that Gisela was not my real mother, but I still had no idea when – much less how or why – she and Hermann had taken me in, and I had pushed the whole business to the back of my mind. I wanted so much to cling to the belief that I belonged to Gisela and her family.

What I couldn't hide from, though, was the way Gisela treated me. She was not cruel; I could never call her that. But she was noticeably cold – emotionally and physically – towards me. This was in stark contrast to her other relationships. Professionally, she was an extremely successful physiotherapist: her patients clearly loved her, and she returned their affection.

With her own relatives, too, she was warm: to her mother and her sister (Aunt Eka, to whom I increasingly turned for love and understanding), and to her son. Hubertus was eight years younger than me; a very handsome boy, who – unlike me at his age – could speak well and fluently. It might have been easy not to like him: he was, after all, Gisela's natural child, and had been living in the house in Hamburg before I was allowed to go there. But although I resented the fact that Gisela seemed able to show love to almost anyone but me,

I had come to care very much for Hubertus and we had a strong bond between us.

But this was a rare glimpse of light. Teenage years are always difficult, especially, I think, for girls. Those crucial years between thirteen and fifteen are generally a time of uncertainty and insecurity, and a time when it is all too easy to be critical of adults. But in Germany in 1956, that biological confusion was exacerbated by national crisis.

The Nazis and the war had broken the previously close bonds of German family life just as surely as the bombs and tanks had destroyed the country's houses, bridges and railways. In addition to creating a huge population of orphans, Hitler's desperate last-ditch battles had blurred the lines between child-hood and adult life by throwing young boys into the doomed fighting.

In the immediate post-war years, an army of international psychologists and social workers was drafted in to address the problems for Germany's next generation. The men and women of the United Nations Refugee Relief Organisation (UNRRA) and its successor, the International Refugee Organisation (IRO), recognised that many teenagers in the late 1940s and early 1950s were growing up without the emotional security they needed – both individually and as part of an emerging new nation. An internal IRO memorandum in May 1949 highlighted the crisis in stark terms: 'The lost identity of individual children is *the* social problem of the day …'

And so, while the American Secretary of State, George Marshall, put in place a vast economic aid plan to rebuild Germany's shattered infrastructure and economy (and the rest of Europe), UNRRA and IRO set to work on what they termed a 'psychological Marshall Plan' for its children.

First they had to identify us. Along with the posters, radio announcements instructed those fostering children from other countries to report to their local youth administration office.

How did this affect us? I could not have told you then what Gisela did: it would be decades before I learned that she met with the investigators without telling me. But when I came face to face with my photo on the poster, I had conflicting emotions.

Of course, I wondered who my real parents were. Perhaps my father had been – like Hermann – an officer in the Wehrmacht, who went away to war, leaving me with a mother who either didn't want me, or could not cope alone with a baby. Those were my rational thoughts. But behind them were the sharp pangs of fear and hope. Hope that my biological mother would see the posters and suddenly turn up to say that she now wanted me. Fear, because if she ever did I was worried what sort of person this woman would turn out to be. Perhaps she would be worse than Gisela; maybe she wouldn't even like me?

But these were only flickering emotions and in the end I found it was easier to snuff them out than to dwell on them. Even though I wasn't happy and I knew that the von Oelhafens and the Andersens were not my blood relatives, I clung to the belief that in some way I belonged to them.

Does it sound odd that the mystery of who I was and where I had come from was never discussed? Perhaps. At the time it was simple: I did not have enough of a relationship with Gisela to ask her difficult questions. It would be a long time before I understood that she might have had good reasons for wanting to leave the past alone.

Whatever the reason, the subject was never broached: to all intents and purposes I was Ingrid von Oelhafen, the name

under which I was registered at school. I did not have one of
the new identity cards, issued by the Federal government from
1951 onwards, but since I was a child no one thought I would
need one until I reached the legal age of majority: twenty-one
in those days.

As it turned out, the problem of my identity surfaced rather
sooner. I wasn't doing well at school: academic work – particularly maths – was not my strong point. I had decided that I
wanted a career as either a children's nurse or a vet, but Gisela
had other ideas. Although she sent me for tests that showed
that I had sufficient potential to take the German equivalent of
A Levels, Gisela wanted me to earn money as soon as possible.
And so it was arranged that I would leave school at the age of
sixteen.

I was unhappy about this decision and felt convinced that
behind everything lay the fact that I was not Gisela's biological
child. But I didn't ask her to change her mind. I made a point
of never asking her for anything because I was afraid she would
refuse. Looking back I think this was a form of self-protection
stemming from the time when I had pleaded in vain with her
to take me away from Hermann's house.

Gisela's plan was for me to train as a physiotherapist, with a
view to at some point coming to work in her practice. But as it
turned out, I couldn't begin the college course in physiotherapy
for another two years. I still have no idea why I was pulled out
of school so early but as a stop-gap, I was sent off to live with
the son of a friend of Gisela's mother, who owned a farm near
Lake Constance on the border between Germany, Switzerland
and Austria. Here I was supposed to learn household management. The farm was in a village called Heiligenholz: it was
remote and tiny, with only three or four houses nearby. For the

first four weeks I cried every night because I was so homesick. Gradually, though, I settled in: the farmer had six daughters and the youngest two, aged twelve and fourteen, became good friends. The farmer's wife was kind and warm, just as I supposed a mother should be. I stayed with them for eleven months and though I didn't really learn any household or cooking skills – my duties were mostly washing up and helping in the fields – they were very good to me, and inadvertently forced Gisela to do something about my lack of official documentation.

At some point during my stay, the family wanted to go on holiday to Switzerland. But I had no papers – no ID card, no passport, not even so much as a birth certificate – which would be needed to cross the border. The only documentation that anyone seemed to have for me was my state health insurance certificate, and that was in the name of the mysterious Erika Matko.

In 1957, children could be included on their parents' documents. Faced with the prospect of leaving me behind (or abandoning his holiday plans altogether), the farmer passed me off as one of his own daughters. We crossed and re-crossed the border without incident. But it prompted him to write to Gisela, urging her to sort out my identity documents – if for no other reason than I was shortly to be dispatched to somewhere where border controls were likely to be less relaxed.

There was still almost a year before I was to begin my physiotherapy training. Rather than spending it back in Hamburg, it was arranged that I was to be sent to England to work as an au pair. I would need a passport.

To this day I have no idea how Gisela arranged it. I never saw a passport in my name, and given what was to follow I'm as sure as can be that I was never issued one. Some form of

documentation must have been procured, however, as I was able
to make the long journey – alone, again – to a small village in
Hertfordshire, 30 miles north of London.

The family I was to live with were evidently wealthy. The
father was a banker who travelled into London every day. His
wife was much younger than him and spent most of her time
with the family's horses. Of their four children, two were away
at Gordonstoun – the famous boarding school where Prince
Charles was a pupil. The third child, an eight-year-old boy,
joined them not long after I arrived, leaving me with only the
couple's five-year-old daughter to look after. I spent six months
in their grand house and thoroughly enjoyed the experience.
The couple treated me very well; I had a lovely bedroom with
a private bathroom, and they made me feel like part of their
family.

Looking back, I doubt I recognised then the irony of finding in the land of my country's former enemy the emotional
warmth I had longed for at home. I was only seventeen and
not as aware of history as I have since become. I returned to
Hamburg with happy memories.

Completely unknown to me, while I had been away the
problem of my identity had once again surfaced. The start date
for my physiotherapy course was approaching and the university needed my birth certificate to register me as a student. I
assume that Gisela was somehow involved in dealing with this.
(It would be many years before I discovered the flurry of correspondence between various local government offices about
me – and in those letters the first hints about my origins.) But
whatever she told the officials was not, I think, wholly truthful.

My new birth certificate – dated 1959, the first time my
existence was formally registered – was in the name of Erika

Matko. It was issued by Standesamt I in Berlin, the Federal government registry which had been specifically created to issue papers for people who had come (or had been forcibly brought) into Germany mainly from the east and who had no other documentation. And yet, oddly, it recorded my birthplace as St Sauerbrunn in Austria. It was a record that would, many years later, hamper the search for my true identity.

At the time, regardless of my birth certificate, I continued to insist that I was Ingrid von Oelhafen. That was the name I answered to and the one by which my friends at university came to know me.

To the university authorities, however, I was someone else: they had registered me in the name of Erika Matko, and when I graduated three years later, aged twenty-one, that was the name on my degree certificate. When I asked the university to change this to Ingrid von Oelhafen, my request was refused: without any official paperwork to prove that I was Ingrid, the administration insisted that I was Erika.

It was 1962 and I was now an adult, about to enter the world of work (and to pay my taxes and social security contributions) for the first time. My first job was in an institute in the Black Forest. I was quite used to being away from home by now: I had not really lived with Gisela and her family in Blumenstrasse for any length of time since I left school. And I found myself thoroughly enjoying my new life away from her and without the complications that had dogged my existence back in Hamburg. It was not to last.

In the last year of my training, shortly before my twenty-first birthday, Gisela had a serious accident. She had fallen down the stairs and lapsed into a coma, which lasted six months. Even when she eventually woke, she was so severely ill that she remained in hospital for another year. My grandmother and Aunt Eka had taken charge of her affairs while she was in hospital but when it was time for her to be discharged, I was needed. With much regret, I left my job in the Black Forest and returned to Hamburg.

Gisela was then forty-nine years old: still relatively young and the mother of a young boy, but now severely disabled. The fall had left her with brain damage and she was quite unable to walk. There was no prospect of her picking up the reins of her physiotherapy practice again: instead it was decided that I would have to do so.

It was the last thing I wanted to do. I felt uncomfortable about taking over Gisela's business and it meant I had to abandon plans to go to America, where I had wanted to study a new technique. I was conscious, also, that the tangled relationships in the von Oelhafen family would not be easy to manage.

I moved back into my old room in the house in Blumenstrasse. It was a difficult time: my grandmother, Aunt Eka, Hubertus and I all had to adjust to our new circumstances – and to Gisela's condition. It was particularly hard for Hubertus to see his mother so disabled, but as she learned to walk in the garden and even to share occasional laughter, his unhappiness eased.

The big problem was that Hubertus and Eka didn't understand one another, and this often led to rows. I was caught between them, which I hated, because each sought my support against the other. Nor were things easy with Gisela. To some extent, she was like a small child and my aunt tended to speak to her as a strict teacher would address a recalcitrant pupil: understandably my mother resented this and became stubborn.

Hubertus and I found it easier to accept her as she now was, though I often found myself feeling aggrieved: I was still made to feel very much like an outsider, with few rights but enormous responsibility. It didn't seem right or fair, but I knew that I had no choice but to get on with the job and make the best of it.

After a while, for the first time in as long as I could remember, my relationship with Gisela improved. We didn't talk much, but she lost some of the coldness towards me that had marred my younger years. I realised, of course, why this was: her disability had softened her and as she was now increasingly dependent on me, she let me see that I was needed.

In other circumstances or other families, it might have opened the door to an open and frank discussion about my past, and how I came to be fostered by her and Hermann. But that never happened. We never spoke about Hermann: I think that after his death she, like me (although for very different reasons), felt free of him and of the burden of their failed marriage. Perhaps because they had never divorced, she had felt tied to him and haunted by the need to justify her refusal to live with him.

Or so I suppose now. Gisela never discussed her marriage with me, just as she never talked to me about my origins.

The question of who I really was had not, of course, gone away. In the mid-1960s I decided to take matters into my own hands. Although I called myself Ingrid von Oelhafen, I was still officially Erika Matko. I felt that the time had come formally to change my name by the equivalent of deed poll.

But the process turned out to be more difficult than I had imagined. I discovered that German law required me to seek the permission of the von Oelhafen family. Even if Gisela had been well enough to do this, the regulations did not recognise her as a von Oelhafen. She had married into the name: the law recognised only those who had been born into it as the true owners of its heritage. Once again, the old German belief in the sanctity of blood resurfaced.

Ironically, Hubertus was registered as a von Oelhafen and so, in theory, allowed to grant permission. But he was legally a child and too young to sign any official documents. I don't know why but Hubertus had an official guardian, a lawyer, and so I had to write to this guardian to plead my case. Eventually he agreed, with one qualification: I was not permitted to call myself Ingrid von Oelhafen, since I wasn't part of the family by blood. Instead I could style myself 'Ingrid Matko-von-Oelhafen' – a signal to the outside world that I was, in some way, a lesser member of the clan. It was hurtful, but there was nothing to be done: I signed the papers and obtained my new name. The certificate cost 100 marks.

I needed to apply for a passport around the same time and was alarmed to discover when I did so that the authorities wanted to classify me as 'stateless'. Apparently the unresolved question of where I had been born, and to whom, was still an obstacle to being recognised as a genuine German citizen. I was stunned: the ruling made me feel worthless, as though I were a nothing, a nobody. Nor could I understand why it should be – after all, I had been paying taxes for more than three years. Worse still, the classification could have meant that I was not allowed to vote in elections and that I would not be able to travel freely abroad.

It took many months, and the assistance of a lawyer friend of Eka's, before the government relented and I was issued with a passport showing me to be a real German. I didn't know it then, but had Gisela been less disabled (or had she been honest with me years earlier) she could have given me a document that she had kept hidden away for more than twenty years which would have cut though all the red tape and bureaucracy. But she wasn't, and she hadn't been – and it would be another three decades before I found it in a cache of other vital paperwork.

Of those decades there is little to tell that is truly relevant to this story. I spent six years in Gisela's house in Hamburg, running her physiotherapy practice. I wasn't particularly happy with this arrangement: Gisela's clientele was largely made up of the elderly, and my interests lay elsewhere. On one occasion a three-year-old girl who couldn't walk came to the surgery: much as I wanted to help her, I wasn't qualified to do so. From that moment I knew that I wanted to work with children.

I found out about a course at Innsbruck, in the Austrian Tyrol, to learn a new technique for helping disabled youngsters. It would take me away from Hamburg for ten weeks, and a locum would be needed to look after Gisela's practice (as I still thought of it) while I was away. Aunt Eka was not happy about me going, but I was determined.

Towards the end of the course, I was offered a job on the staff at Innsbruck University clinic: I worried about what my aunt would say – and what would happen to Gisela – but in the end I accepted and spent a happy year doing what I loved in a place where I felt comfortable.

It was during my time in Innsbruck that I fell in love. I met a young man who came from Osnabrück – close to Bad Salzuflen, where I had lived with Hermann, and even closer to the house in Hamburg. We began a life together in Osnabrück, albeit in separate apartments.

Our relationship didn't last. For whatever reason, I did not find – have never found – it easy to maintain adult relationships with men. Whether this has something to do with my childhood I cannot say: I know only that whilst I like the idea of falling in love, those to whom I have often been attracted don't feel the same way about me, and I never seemed to like those men who did see something in me that appealed to them.

I do not say any of this to elicit your sympathy: I am not, by nature, comfortable with that, and if I have never known the intimacy of married life or had my own children, I have been fortunate always to find and keep good friendships with other women which have sustained me. And I have known, too, the joy of helping many, many children: in the early 1970s, after years of working in hospitals, I established my own physio-therapy practice dedicated to working with disabled youngsters. From then on I worked six days a week, twelve hours a day, consumed by the need to help them. Every year I travelled across Europe, England and the United States, attending spe-cialist courses that extended my understanding and developed my skills. It has been a lifelong and rewarding career that has brought me immense happiness.

But what of Gisela, the von Oelhafens and the Andersens? What of the life I stepped away from in Hamburg, the strange mystery of my birth and the circumstances by which I came to be fostered? Although I remained in touch with the family throughout my adult life – and stayed close to Eka, in particu-lar – I did not return to live with them, nor work in Gisela's business.

By the time I did – briefly – go back, the walls had come down across Germany and the east.

## SIX | WALLS

*'Because she is a child of German stock,
on the orders of the Reichsführer she is
to be brought up in a German family'*
STURMBANFÜHRER GÜNTHER TESCH

At 10.45 p.m. on Thursday, 9 November 1989, the Berlin Wall – that most visible and entrenched symbol of the Iron Curtain – began to crumble. I was forty-eight years old: for almost half a century my life, and the lives of my compatriots, had been shaped by the division of our country into East and West.

It had been a bitter partition: beyond the wall East Germany had re-enforced its borders, imprisoning its population inside a rigid ideological police state. Those who sought to flee, as Gisela had done with Dietmar and me, regularly found their way blocked by barbed wire and checkpoints – and by troops under orders to shoot anyone trying to make their way to freedom. More than a thousand men, women and children had been killed trying to escape the iron grip of communism.

And then it was over. After a day of confusion and rumour, the East German commander of the key border checkpoint opened the gate and ordered his guards to allow people through.

Hundreds of *Ossis* – as those from the East were known – swarmed through to be greeted by West Berliners waiting with flowers and champagne.

Before long, a crowd of *Wessis* climbed on top of the wall, where they were joined by East German youngsters. They danced together, joyously celebrating their new freedom. Within hours, television cameras captured images of people using hammers and chisels to chip lumps off the wall: soon these *Mauerspechte* (literally 'wall woodpeckers') demolished entire lengths of it, creating several unofficial border crossings.

The speed of events took the governments of both East and West Germany by surprise. Yet, in truth, change had been in the air for months. It began in August when Hungary – physically and politically one of the outliers of the Eastern bloc Moscow had created – effectively dismantled its physical border with Austria. Within weeks more than 13,000 *Ossis* had travelled to Hungary and then on into Austria. When the government in Budapest tried to stop the flow, East Germans simply marched into the West German embassy and refused to return home. It was an unprecedented act of civil disobedience from a nation which had, for fifty years, grown used to obeying the orders of its communist masters, and there was more to come.

Throughout that early autumn, mass demonstrations broke out across East Germany. Protestors took to the streets chanting *'Wir wollen raus!'* ('We want out!') and *'Wir sind das Volk!'* ('We are the people!'). Newspapers and television stations began proclaiming the dawn of a peaceful revolution.

By the time Erich Honecker resigned as General Secretary of the ruling Socialist Unity Party in October, the movement was plainly unstoppable. Honecker was not merely the head of state: as the man who had been in control of East Germany since the early 1970s, he was seen as the embodiment of the communist state itself.

Despite the warning signs, the collapse of the physical borders was chaotic and unplanned. In the early afternoon of 9 November, a televised press conference in East Berlin first hinted that a limited exodus might be permitted. But after hearing the broadcast, people began gathering at the six check-points between East and West Berlin, demanding that border guards open the gates immediately. The soldiers were taken by surprise and overwhelmed by the sheer number of those seeking to cross into the West: they made panicky phone calls demanding instructions.

It soon became clear that no one in the disintegrating East German government would take personal responsibility for issuing shoot-to-kill orders: as a result the border guards simply stepped aside and allowed the huge crowds to pass peacefully into the West. At a little before 11 p.m., West German television pronounced the last rites of the German Democratic Republic.

> This is a historic day. East Germany has announced that, starting immediately, its borders are open to everyone. The GDR is opening its borders ... the gates in the Berlin Wall stand open.

The opening – and the determined dismantling – of the Berlin Wall was followed inevitably by the abandoning of all check-points between East and West Germany. By 1 July 1990, the day the deutschmark was adopted throughout Germany, all border controls officially ceased to exist. Three months later, East Germany was dissolved and absorbed into a new unified Republic.

What did all this mean to me? Although I was born in the early years of the war, I was really a child of the 1950s and 60s – decades in which West Germany had sought to hide the crimes of the past amid the divisions and troubles of the present. I cannot pretend that the reunification of my country meant any more to me than it did to most people of my generation: we were thankful to have grown up on the 'right' side of the Iron Curtain and vaguely reassured that the tide of history had somehow restored the proper and natural order. To be sure, there were economic concerns: no one seemed quite certain what the cost of our new country might be, though there were regular and dire predictions that the German economic miracle, for so long the envy of Europe, would be threatened by the need to support our less developed, and bankrupt, former neighbour. But such fears were primarily for politicians, less alarming for a physiotherapist with her own successful practice living in the security of Lower Saxony. When reunification happened I was fifty years old. I had never married and my life was comfortable: I was financially secure, had a lovely home and I was working harder than ever. There were, though, clouds building. And, inevitably, they centred around Gisela.

My foster mother's health had worsened as the years had passed and she was now severely disabled. Tragedy had also struck the family. Hubertus, the handsome little boy I had known as a child, had grown up into an attractive gay man: in the mid-1980s he was among the first German men to be diagnosed with the terrifying – and then always fatal – new disease of AIDS. In 1988 it claimed him.

The decision to hire a full-time carer to look after Gisela seemed the best way of securing her future. Gisela was well off. Her practice had provided well for her, and both the von Oelhafens and Andersens had money. But the woman we hired saw an opportunity. Not long after Hubertus's death, she took advantage of Gisela's grief and enfeebled mind and persuaded her to move to Gran Canaria, where, she said, they would benefit from a warmer climate. And so the two of them set up home together, more than 3,000 miles away from any of Gisela's relations. Worse, her companion worked hard to cut us all off from Gisela and to isolate her: none of us was able or allowed to contact her.

Only when Gisela developed dementia was I permitted to visit her. What I found in Gran Canaria disturbed me greatly: it was apparent that Gisela was entirely dependent on a woman whose chief concern was to extract as much money as possible from her before she died. Something had to be done.

Together with Aunt Eka I petitioned the German Guardianship Court to order that we be allowed to intervene in Gisela's life. At first the court refused to hear our plea, saying I had no claim because I was only Gisela's foster daughter, not her biological child. But unusually for me (I am not, by nature, forceful), I dug my heels in. I said to the judges: 'I will sit here until you listen to me. I will stay here in this court until you listen to what I have to say.'

Eventually they agreed to hear me. I told the court that Gisela was being controlled by her carer and that she had manipulated the relationship to such an extent that she had been named as one of the main beneficiaries of Gisela's will. I begged them to safeguard her interests. But listening to me was as far as the judges were prepared to go. Ultimately, the court declined to intervene.

It was left to Aunt Eka to work out a private compromise settlement, which provided some measure of protection for

Gisela. But the damage had been done: Gisela lived on until 2002, but never again would we be a family.

Her exile in Gran Canaria did have one positive outcome. When Aunt Eka and I finally realised that Gisela would never come back to Hamburg, we set about clearing out her rooms. Which is how I came to find the diary she had kept of my earliest years.

I will remember for ever the moment I laid my hands on it, and the emotion I felt reading its few handwritten pages. I was so very thankful: I had found something about me and my early life – it was the first time I could reach out and touch my past. But alongside the joy there was pain too.

I think, perhaps, I hadn't realised the extent to which I had for nearly forty years blocked off my feelings about the mystery of my childhood. Holding the little volume, the sense of loss and uncertainty was overwhelming. Why had she not given me this diary but instead kept it hidden? How could she not have realised what it would mean to me?

What made this all the more painful was the knowledge that I had only discovered the book because Gisela had – to all intents and purposes – abandoned me once again. That she was in no state to understand this, and that her carer was deliberately exploiting her frailty, did not change the fact that I could not contact her to ask all the questions which the diary prompted.

Perhaps my overpowering sense of loss and hurt explains why I did not look more closely at the other documents I found in Gisela's room. I glanced at them and saw that they seemed to be legal papers about the process by which Gisela and Hermann had fostered me. But rather than pay them the attention I should have, I put them away and devoted myself to my work. It wasn't until the end of the twentieth century that I was reminded of their existence.

∽∞∾

One day in the autumn of 1999, I was at my practice as usual when the phone rang. I assumed the caller was a patient or perhaps a referral for a new client. But the lady on the phone that morning was neither of these things. She first asked whether I was Ingrid von Oelhafen and then explained that she was from the German Red Cross. I was initially puzzled: why would the Red Cross be ringing me? I had no professional connection with the organisation: certainly none of my patients had ever come from there.

Then she asked a question that took me completely by surprise: would I be interested in looking for my birth parents?

I find it hard to describe the feelings that ran through me in that moment. For so long I had put the questions regarding who I was and where I had come from to the back of my mind, telling myself that working with disabled children was more important; in truth, though, I think I was really avoiding the issue, perhaps for fear of what I might find. And so I was surprised to find that my overriding emotion was one of real excitement: at long last I had the chance to find out about my origins. Perhaps I was now finally ready to face the truth.

I have thought about this a great deal and come to the conclusion that it was age that made the difference. I was fifty-eight when I received that phone call, and looking back now I can see that the older I became, the more I wondered about my personal history. I am not alone in this: it is part of the human condition to revisit the past as the years slip away. There were practical considerations, too. Whenever I had cause to go to the doctor – something that becomes more frequent with advancing years – I was asked about my family medical history, and of course I had to say that I had no idea.

I didn't ask how the Red Cross knew where to find me or how they knew that I had a family mystery to solve. I simply said yes, and hoped for the best. They couldn't give me any concrete information about my past. Instead, the woman told me to contact an academic historian at the university in Mainz.

I owe an immense debt to Georg Lilienthal. When I sat down to write to him, I had no idea who he was – much less how important his role would be in my story. I simply knew what the Red Cross had told me: he was the person who could set my feet on the path I would need to follow.

I understood that Dr Lilienthal would be expecting my letter, so I wrote openly and honestly, explaining that I had always wanted to know where I came from but that I had never known how or where to start.

When I posted the letter I was so excited I wanted to get in the car and drive to Mainz the next day. But something told me that I must wait: whatever information this man had, he would surely need time to pull it together. And so I resolved to be patient and use the time to search through the documents I had found in Gisela's room. I felt tantalisingly close to discovering the story of how I came to be fostered by Gisela and Hermann and it was frustrating to still be in the dark. But I had waited fifty years before embarking on this quest: a few more weeks wouldn't kill me.

I dug out the box of papers. In the years since I had found them I'd never even taken a look at anything other than the diary. Now I began to look closely at the sheaf of fading documents Gisela had kept with it.

The first was a small and slightly dog-eared pink slip. It was a vaccination certificate, dated 19 January 1944 and signed at Kohren-Sahlis, near Leipzig: it showed that Erika Matko, born on 11 November 1941 in a place called St Sauerbrunn, had been inoculated against scarlet fever and diphtheria.

The date was significant: January 1944 was several months before I had been fostered by Gisela and Hermann. But other than indicating that the signatory was a doctor, nothing else on the form showed where the vaccination had taken place, or at whose request. What organisation had been based at Kohren-Sahlis? And, for that matter, where exactly was St Sauerbrunn? A second certificate recorded further vaccinations. On the reverse side was an official stamp that read: *'Lebensborn Heim Sonnenwiese Kohren-Sahlis'*.

*Heim* meant a children's home: that much I knew from my earliest days, and it certainly fitted with Herman and Gisela having fostered me. But what was Lebensborn? I had never heard the word before.

The next document was even more puzzling. Dated 4 August 1944, it appeared to be a kind of contract-cum-receipt for my foster parents.

> The family Hermann von Oelhafen, of Gentz Strasse 5, Munich, has on 3 June 1944, taken into their home the ethnic German girl Erika Matkow [sic], born 11 November 1941. Because she is a child of German stock, on the orders of the Reichsführer she is to be brought up in a German family.
>
> There will be no provision for the maintenance of

the child from either side: the child herself has no assets
or revenue. The foster parents alone shall be responsible
for her support.

The certificate had apparently been issued in Steinhöring. This,
I knew, was a small village not far away from Munich, but
there was no other information about the organisation that had
created it. The only clue was the letterhead at the top of the
paper, almost obscured by holepunch holes and the passage of
time: *'Der Reichskommissar für die Festigung deutschen Volkstums,
Stabshauptamt L'*. I had no idea what this could be: a little
research revealed it to be the office of the Reichs Commissioner
for Strengthening German Nationhood, a Nazi organisation.
What exactly the office did was not immediately clear.

At the bottom of the document was the signature of a
Doctor Tesch, who described himself as a Sturmbannführer.
Anyone who had grown up in Germany after the war knew
that word: it was a paramilitary rank in the Third Reich,
equivalent to a major in the regular army but almost exclusively reserved for use by members of the SS. Why would an
officer in Heinrich Himmler's reviled Death's Head* organisation have had anything to do with my foster care? I looked
again at the certificate: it said that I had been handed over to a
German family 'on the orders of the Reichsführer'. That was
Himmler again. Bafflingly, it looked as though Hitler's second-
in-command, the most feared man in Nazi Germany, had played
some kind of role in my childhood.

---

\* The SS cap badge was a Death's Head (skull and crossbones): because
of this and their role in administering the death camps, they were known
as SS-Totenkopfverbände – literally, Death's Head units or regiments.

I was desperate to ask Gisela what all this meant – and, indeed, why she had kept these documents from me for so many years. But Gisela was in Gran Canaria and, by this stage, in the last throes of her dementia. I knew I would get no help from her.

A week had now passed since I sent Georg Lilienthal my letter: I wondered if he was away from his office or whether he was for some reason unwilling to share with me what the Red Cross said he knew – or at least suspected – about my history. In the interim, I decided to begin my own investigations. I wrote to the German state archives (the Bundesarchiv) to ask if they held any documents bearing my name or that of Erika Matko.

I assumed, naively, that the Bundesarchiv would reply quickly: how difficult could it be, in this age of computerised databases, to run a simple check on my names? I was about to discover one of the paradoxes of the new Germany: while the new state was committed to uncovering the terrible sins committed by the rulers of the old East German state, and zealous in rooting out of public life those who had been involved with its secret police, the Stasi, it was much less willing to face up to the crimes committed by Hitler's Thousand Year Reich.

In part this was a legacy of the early post-war years. Konrad Adenauer, West Germany's first chancellor, had vehemently opposed much of the Allied powers' work on denazification, and had pushed for the release of those convicted of war crimes at the Nuremberg trials. He had even appointed, as his right-hand man in government, Hans Globke: a politician who had drafted anti-Semitic laws for Hitler in 1938.

From the outset, no one wanted to look too closely at the past, and many years later at the end of the twentieth century, despite its proud position as the driving force of the European

Union, Germany still had skeletons in its historical cupboard – skeletons it was neither ready nor willing to rattle.

The Berlin Wall had not been the only barrier separating Germany from itself. If the nation was now reunited, our collective memory was still decidedly patchy. Over the coming months I would discover that anything relating to the mysterious Lebensborn programme seemed to spark repeated bouts of amnesia. There had been very little published about it, and what information was available suggested a story of national shame and a legacy still shrouded in secrecy.

As I waited for Georg Lilienthal and the Bundesarchiv to respond to me, I thought back to the telephone call from the Red Cross. The woman had seemed reluctant to give me any information: had she been trying to warn me about the problems I would face when she asked whether I really wanted to investigate my past? Perhaps, but however difficult the task might be, I was determined to try. I did not realise then, as I tentatively began my personal quest, that I would also be embarking on a painful journey into Germany's troubled history, as well as that of a country it had once invaded and plundered.

*'The eternal law of nature to keep the race pure is
the legacy that the National Socialist movement has
bestowed upon the German people for all time.'*

NAZI PROPAGANDA FILM, 1935

There was no such place as St Sauerbrunn.

With little else to go on, I returned to the very first record of my existence: the little pink slip of paper showing that I had been vaccinated against scarlet fever and diphtheria. As it documented my birthplace as St Sauerbrunn, that seemed the most logical place to start. But although I searched through atlases and historic maps of Germany and all the countries Hitler's armies had invaded, there was no town or village with that name.

The closest match was the Austrian spa town of Bad Sauerbrunn, close to the border with Hungary. At the start of 2000, I found the address of the Austrian Ministry of Foreign Affairs and wrote a lengthy letter asking if they could help me locate any record of a family called Matko anywhere in the vicinity of Bad Sauerbrunn.

I was now becoming a little impatient. I had not received a reply – much less any information – from the Bundesarchiv, and I was still waiting for Georg Lilienthal to deliver the information he had supposedly found about me during his research into Lebensborn.

Frustrated, I began searching for information about this mysterious-sounding organisation. What struck me immediately was how little seemed to have been published. More than fifty years after the end of the war, the terrible history of the Third Reich and its crimes had been analysed and picked over in meticulous detail, and yet a Google search for Lebensborn produced only a few, largely repetitive results.

Ostensibly, the Lebensborn Society (literally translated, *Lebensborn* means Fount, or Source, of Life) was founded in 1935 as some sort of welfare organisation, funded by the Nazi Party, to run maternity homes across Germany; it was set up in response to what was rapidly becoming a demographic crisis for the new Reich. When Hitler came to power in the 1930s, the country's population had been falling for decades. In 1900, the statistics showed an average rate of births per thousand of 35.8; by 1932 that had dropped to 14.7. From the outset, the Nazi regime set out to stop and then reverse the trend. They began with slogans – 'Restoring the family to its rightful place' was typical – and then introduced financial incentives such as marriage loans, child subsidies and family allowances to encourage large families. A cult of motherhood was also formally established: every year on the birthday of Hitler's own mother, fertile women were awarded the Honour Cross of the German Mother. Those who produced more than four children were given a bronze medal; more than six earned silver; and gold was awarded to those with more than eight.

When this didn't produce results quickly enough, new laws were introduced to ban the advertisement and display of contraceptives and Germany's pioneering birth control clinics were shut down (in the 1920s Germany had been a world leader in developing contraceptive devices such as the IUD). Abortions were criminalised as 'acts of sabotage against Germany's racial future'.

That phrase, 'racial future', was my first clue to the reality hiding behind the seemingly innocuous Lebensborn Society. Although the ostensible aim of the homes was to allow women who might otherwise abort their pregnancy to give birth in safety and in secret – thus helping to boost Germany's population – they weren't open to everyone.

I was, of course, aware of the Nazis' obsession with race: it was the altar on which Hitler and his regime had sacrificed more than six million Jews. What I hadn't encountered was the extraordinary and convoluted web of organisations that had been established to safeguard the 'purity' of the German race. As I continued my research, I could feel myself being pulled down the rabbit hole of National Socialist madness. At its heart was the sinister figure of Heinrich Himmler.

Himmler had joined the Nazi Party in August 1923, three years after its birth. He was not one of its early fanatics – his membership number was 14,303 – but within six years he had taken charge of its most powerful paramilitary organisation, the Schutzstaffel, more infamously known by its initials: SS.

As Reichsführer-SS, Himmler began creating a parallel, and ultimately much more powerful, organisation to control and monitor the Nazi Party. He had long been interested in the then-fashionable quasi-science of eugenics and became obsessed with the idea of a mystical past in which a Nordic race of pure-blooded warriors had conquered much of Europe. He began reorganising the SS to be the vanguard of a reborn race of Aryan 'supermen'. Under his direction applicants were vetted for their racial 'qualities': he described the process as being 'like a nursery gardener trying to reproduce a good old strain which has been adulterated and debased; we started from the principles of plant selection and then proceeded quite unashamedly to weed out the men whom we did not think we could use for the build up of the SS'.

In 1931, he created a separate department within the SS to ensure his 'plant selection' ran smoothly: Das Rasse-und-Siedlungshauptamt-SS, or RuSHA. A literal translation would be 'SS Race and Settlement Main Office'; what it meant in practice was an organisation dedicated to safeguarding the 'racial purity' of the Schutzstaffel. One of its duties was to oversee the marriages of SS personnel: on Himmler's personal orders, RuSHA only issued a permit to marry after detailed background investigations had proved that both partners had an uninterrupted racial pedigree showing them to have come from pure Aryan blood-stock as far back as 1800.

As I read further, I discovered that the Lebensborn Society had been formed and fostered under RuSHA's banner. In a circular issued on 13 December 1936, Himmler had set out both the lineage and the aims of his new organisation:

The Lebensborn Society is under the direct personal control of the Reichsführer-SS. It is an integral part of the Race and Settlement Head Office and its objects are:

1.   To support racially and genetically valuable large families.

2.   To accommodate and look after racially and genetically valuable expectant mothers who, after careful

investigation of their families and those of the fathers of
their children by RuSHA, can be expected to give birth
to equally valuable children.

3.  To look after those children.

4.  To look after the mothers of those children.

Even to me – a German woman, born during the war, who had
lived her whole life in a country trying to come to terms with
the legacy of Hitler's twisted vision – this sounded the stuff of
madness. In German we have a very expressive word for this
sort of fantastic lunacy: *unglaublich*, meaning unbelievable. How
could anyone 'prove' their racial or genetic value – and what in
any event did such a bizarre concept mean in practice?

I was not, it emerged, alone in being confused. I found a
succession of references to rumours which had grown up about
Lebensborn facilities. Some of these dated back to the war years
and suggested that ordinary Germans had become worried by
stories that these ostensible maternity homes were in fact SS
stud farms: places where the cream of Himmler's brigades were
introduced to suitable Aryan women in order to breed racially
valuable babies for the Reich. The gossip was false but the
secrecy that surrounded Lebensborn ensured that the rumours
had persisted over the years. There was even an entire genre of
Nazi exploitation films and books dedicated to mythologising
the programme: one typical example, a movie made in 1961 by
a German director and widely available on the Internet, had
the English title 'Ordered to Love' and the screaming subtitle
'Frauleins Forced into Nazi Breeding!'

I felt ashamed and horrified. I understood that the SS stud
farm stories were no more than absurd fantasies (and as often

as not, cynical attempts to sell tawdry films and novels), but if
this was what the world knew – or thought it knew – about
Lebensborn, was it any surprise that modern Germany was
unwilling to talk openly about it? Perhaps this explained why my
requests for information from the Bundesarchiv were still unan-
swered, two months on. Alone, I had little chance of being able
to investigate Lebensborn, much less to uncover its role in my
origins. I needed help, but there seemed to be a wall of silence
surrounding anything to do with this corner of Nazi history.

∞

In February 2000, my hopes were dashed further. The Austrian
government finally replied to my letter about Matko family
records in Bad Sauerbrunn: there was no such record and never
had been. My journey seemed to be over almost before it had
started. If I hadn't come from Austria, where had I been born?

And then, a few days later, a letter arrived from Georg
Lilienthal in Mainz. For the first time it contained clues – solid,
historical information – about Lebensborn and how it fitted
into my own story. He wrote cautiously, hinting once again at
painful secrets lying in wait for me.

Dear Frau von Oelhafen,

I would first like to thank you for the trust you have
placed in me with your letter because it's all about the
question of your identity. Therefore, I am also glad that
Ms Fischer of the German Red Cross in her conversation
with you was very careful … I have to apologise to you.
My response to your letter took a long time. And while

you were waiting for a sign from me, you might have come to doubt whether your request was the right thing to do. I can reassure you.

My long silence was partly due to external reasons (too little time to find the documents together and write): but on the other hand I was also aware that the answer would not be easy because I know what it could mean to you. That's why I have been writing my letter on and off since early January. That is what has led me to outline your presumptive fate so soberly and in a seemingly emotionless way. I did not want to influence your feelings with my feelings.

Now for your request. As you write, the fact that you have two names (Erika Matko and Ingrid von Oelhafen) has long been known. I assume that you have therefore always wondered what it's all about. Apparently your foster parents were not completely open with the little they knew about you.

After reading the letter, I could not have told you exactly what my feelings were: there was anxiety and apprehension, but these were also shot through with excitement.

I knew, of course, that neither Hermann nor Gisela had ever told me the truth about my origins. I had, to some extent, convinced myself that if this hadn't been the result of the tensions of post-war life, it must have been because they didn't really know my history. Lilienthal's letter was the first time I had to confront the possibility that my foster parents might have deliberately withheld information.

And then came the revelation I had been waiting for and half-expecting. Lilienthal's research had found the name Erika

Matko in some long-forgotten records of the Lebensborn programme. She had been raised in one of its children's homes:
a place called Sonnenwiese (literally: 'Sunny Meadow') at
Kohren-Sahlis. Since I was – or once had been – Erika Matko,
that meant I was a Lebensborn baby. What's more, his investigations had convinced him that Hermann and Gisela had
deliberately concealed this information from me.

There was now solid documentary evidence linking me to
this bizarre and evidently still-shameful Nazi organisation: one
which had been under the direct control of the SS. And yet my
overall response was exhilaration rather than shock. It seemed
incredible, but it also offered the chance finally to find out
more about who I was and where I had come from. And in a
way, the revelation also brought me a little peace. Although I
did not yet understand the true nature of Lebensborn, I could
now let go of one of the worries which I had lived with ever
since I discovered that I had been fostered.

If, as my initial research indicated, Lebensborn was a political programme in which the demands of the Nazi regime
overrode the feelings of those it ruled, perhaps the reason my
real parents had given me up was also political, not (as I had
feared) the much more upsetting idea that they simply hadn't
wanted to keep me. That realisation brought me some comfort,
but also a hint of fear. In some way I had been involved with an
organisation which, nearly sixty years later, still evoked fear and
loathing. I mentally added the SS to the growing list of Nazi
groups I would need to investigate.

Lilienthal's letter contained more surprises. Realising that I
knew little about how Lebensborn had operated, he explained
how children had arrived at Sonnenwiese. Some had been born
in Lebensborn maternity homes, then brought to Kohren-Sahlis

as part of Himmler's programme to increase the population of the Reich. Others, though, had apparently been kidnapped.

> Children born in this maternity home were German children born illegitimately in the Lebensborn programme for foster care or for adoption. But there were also children at Kohren-Sahlis who were trafficked from the occupied countries of Germany and who were designated for Germanisation.

I had never heard of 'Germanisation'. Why would the Nazis traffic children from the countries they had invaded? I had always been taught that Hitler and his henchmen viewed the people of many of these conquered states quite literally as 'sub-human'. And how did this fit with my background?

> Lebensborn worked with German foster families, with the intention of later adoption after the victorious end of the war. The fall of the Third Reich prevented these plans from being realised. Most of these foreign children returned to their home countries. However, some remained in Germany with their foster families.
>
> There were various different reasons for this. Some of the foster parents cared for their foster families. Some of the foster parents concealed the foreign origin, even from the children themselves, for fear that the children might be removed again or that they would have a yearning to return home. Ultimately, they were afraid to lose the love and affection of their foster children. Also, they wanted to protect the children from hostility and integration difficulties.

These were often the reasons that the children were not adopted after the war, as well as the fact that they often lacked the necessary papers.

Some of the Allies did not want to send the children back to their home countries against their will, and they remained in the German family with the authorities of their home countries in agreement because they had no biological family left.

And then Lilienthal dropped his biggest bombshell.

Frau von Oelhafen, is it your belief that you might not be a child of German parents? I have known your name 'Erika Matko', and the name of your foster parents 'von Oelhafen', for many years from documents in the Bundesarchiv. I have researched Lebensborn for over twenty years and I know many of the fates of Lebensborn children.

Their names are mentioned in lists that were created by Lebensborn for children to be Germanised from Poland, Yugoslavia and Czechoslovakia (in the Lebensborn management they were called only *Ost-Kinder*) and in the records and statements of former Lebensborn employees.

Although I can present to you no papers (such as a birth certificate), giving you ultimate peace of mind, I do have documents that seem to show that you probably originated in Yugoslavia.

After you have read my letter, you may ask: what to do now? I cannot give you the answer to this. But if you want to continue to search for your identity, I will be happy to assist you. You can always contact me.

Kidnapping, Germanisation, *Ost-Kinder*: these words and notions were so alien to me, so far away from the assumptions I had made when I began my investigation, that I didn't know what to make of them. Although the Austrian authorities had been unable to find any trace of a Matko family in or near Bad Sauerbrunn, I had still believed that the search for my past would somehow take me to Austria. It had been a comforting thought in a way: the time I had spent at Innsbruck meant the country felt familiar. Nor was there any language barrier: German was the national language. Now it seemed I had to start again from square one – and in a language I had never even heard spoken. Even worse, Yugoslavia itself had ceased to exist: the last of the former Iron Curtain countries had disintegrated in a bloody civil war before splintering into a series of smaller new states. Where – how – would I begin?

I decided to take Georg Lilienthal at his word. I wrote to him, asking for guidance. Throughout my journey into the past I have been very fortunate to find people who were willing to give their time and share their expertise to assist me from one stumbling step to the next. Dr Lilienthal was the first, and probably the most important of my guides. He told me I needed to write to two German ministries in Berlin – Foreign Affairs and Internal Affairs.

He helped me compose the letters, each of which explained my situation and set out my belief that I had been brought into the Lebensborn programme from the former Yugoslavia: I requested assistance in making contact with their counterparts in Eastern Europe.

My requests fell on deaf ears. Both ministries sent abrupt and unhelpful replies, saying that they could not do anything for me: the only thing they could suggest was that I write to the government of Slovenia – the new nation that had emerged in the central part of Yugoslavia, once controlled by Hitler's Reich.

Around the same time, I received a reply to my original enquiry at the Bundesarchiv. This too was unhelpful: the state archives insisted that they held nothing relevant to my past. A pattern was developing: no government institution seemed interested in helping me investigate my past. Since I knew that Georg Lilienthal had already found documents relating to Erika Matko and the Sonnenwiese home in those very same archives, it was clear that German officials were reluctant to talk about Lebensborn. Over the next few months it was a reluctance I would encounter over and over again.

Georg Lilienthal pointed me toward two other, lesser-known collections of documents where, he said, I might find information about Lebensborn. And he agreed to use his own contacts to find out who I should write to in Slovenia.

Looking back, I realise that this was the pivotal point in my investigation: from here on there would be no turning back. Once I began digging into boxes of dusty papers, stored in archives across modern Germany, there was no way of knowing what skeletons I might disturb, what secrets I might unearth. Such is the benefit of hindsight. At the time I didn't stop to think about what I was doing: for so long I had avoided thinking about my past, but now I was determined to find out whatever was known about me, and by extension about those who had raised me. If that meant asking questions that made people uncomfortable – well, so be it.

# EIGHT

# BAD AROLSEN

'Adolf Hitler has led the German people to the
realisation that the Nordic race is the most creative,
valuable race on earth. Therefore, caring for the
valuable Nordic blood is their most important task.'

HEINRICH HIMMLER, *RACIAL POLITICS*
(1943 SS PUBLICATION)

Bad Arolsen is a small, picture-postcard German town. For more than 250 years it was owned and ruled by the Princes of Waldeck-Pyrmont, then a sovereign principality stretching across the rich agricultural heartlands of Hesse and Lower Saxony. This aristocratic family constructed a large baroque-style stately home and drew up plans to build the town around it in a mathematically perfect grid of streets. But when they ran out of money, the grandiose scheme was only half-completed: to compensate, the undeveloped sections were landscaped with shrubbery.

Die Grosse Allee is the main street, running one perfectly straight mile from east to west and lined with 880 German oak trees in strict military formation. Exactly halfway down is an unprepossessing piece of post-war architecture, set back from the road behind long hedge walls so as to be almost unnoticeable to the casual visitor. It is the archive of the International Tracing Service. Here, spread haphazardly over several floors and spilling out into satellite buildings, more than thirty million individual files record the fate of those who fell victim to the National Socialist criminal enterprise.

It is a cliché of modern history that the Nazis were painstaking record-keepers. But the 26,000 linear metres of original documents and 232,710 metres of microfilm housed at the ITS bear witness to this thoroughness and, according to Georg Lilienthal, somewhere in the vast piles of paperwork there was probably a record of how I came to be part of the Lebensborn programme.

I wrote to the archive in the early spring of 2000, asking for its help in locating any document that would help me investigate my origins. In theory, this should have been a straightforward request: it was, after all, exactly what ITS was established to do. But, as I was finding, theory and practice remained a long way apart – and what often separated them was politics.

In 1943, the Supreme Allied Headquarters in Europe asked the international section of the British Red Cross to set up a registration and tracing service for missing persons. Even by that mid-stage of the war, Washington and London had begun to plan for its aftermath, conscious that by the end there would be a vast population of the displaced or the disappeared. The Central Tracing Bureau was established in February 1944: as the war shifted eastwards into each territory successively liberated from the German armies, it moved from London to Versailles, then on to Frankfurt before finally arriving at Bad Arolsen in 1946. Here its researchers set about creating an archive of Nazi documents.

The records came from every corner of the former Reich. Allied forces had rescued them from concentration and death camps or captured them from Wehrmacht field offices and Nazi central registries. Each individual piece of paper was analysed: from them the CTB was able to begin reconstructing the fate of tens of millions of men, women and children who had been taken for slave labour, imprisoned, or murdered in the Holocaust.

From the outset the Allies had two, sometimes conflicting, aims for this unprecedented exercise. The first was to prepare reliable documentary evidence for use at the Nuremberg War Crimes Tribunal: for the first time in history, the surviving leaders of a country were to be put on public trial for the newly defined offences of crimes against humanity, conspiracy to wage aggressive war and the industrial-scale murders of Jews and Eastern Europeans (among many others).

The second, longer-term ambition was to create a mechanism to enable the survivors of the war – and especially of the Holocaust – to find their families and, if possible, eventually to reunite them. And so the CTB began building from the captured files a central name index of every single person they could determine to have been a victim of the Nazis' reign of terror.

Whether or not the Allies realised the scale of the task when they began it, they were soon overwhelmed by the sheer volume of cases. The central name index alone would come to house the individual fates of fifty million people. Behind each handwritten index card was a mound of paper.

As the years passed, responsibility for managing and funding this Herculean effort was passed from one organisation to another. In July 1947, the newly formed United Nations' International Refugee Organisation took over administration of the bureau, changing its name to the International Tracing

Service. Less than four years later, it was handed back to the Allied High Commission for Germany – the body set up by America, Britain and France to run their sectors of the former Reich. When the occupied status of Germany was repealed in 1954, ITS was hived off to the International Committee of the Red Cross, which promptly insisted on appointing its own manager to run all daily operations, who, for good measure, had to be a Swiss citizen. It was a sorry catalogue of financial and administrative buck-passing which ensured that the ITS was destined to become the Cinderella of the vast post-war archives mission.

The situation worsened with the implementation in 1955 of the Bonn Agreement, which formally ratified the new nation of West Germany. One clause in this document prohibited the publication of any data that could harm former victims of the Nazis or their families. However well intentioned, the instruction effectively shut the Bad Arolsen archive off from public scrutiny: historians and journalists were not permitted to examine its contents, and although individual victims of the tyranny were theoretically able to ask for any relevant information, this too became caught up in the *Realpolitik* of modern Europe.

At the start of 2000 – just as I made my request for help – the German parliament was under pressure to set up a fund to compensate an estimated one million survivors of the Nazis' forced and slave labour programme. These were men and women who had been shipped from Eastern Europe to toil in the factories that kept Hitler's war machine running. Soon the Bundestag passed a law establishing a Remembrance, Responsibility and Future Foundation (*Stiftung Erinnerung, Verantwortung und Zukunft*) that would make payments to those who could prove they had been affected. The evidence they needed was primarily held at the ITS: it was almost instantly flooded with applications, and all other enquiries were either ignored or not properly processed. Among them was my letter: I received a brief, and as it would turn out completely inaccurate, reply to the effect that there was no trace of me in the files.

Seven years would pass before the ITS archives were opened to full public scrutiny: lost time that would have a terrible impact on the search for my biological family. But to explain the origins of Lebensborn, I need to step away from my own chronology to pull back the veil of secrecy which then surrounded Bad Arolsen. Among the millions of documents captured from the Nazi war machine were many of Heinrich Himmler's personal papers. These were sent to the ITS, where separate folders were opened, each covering the myriad organisations the Reichsführer had set up, as well as the bizarre and obsessive belief system that underpinned them.

The pernicious idea that one race was superior to another by virtue of the purity of its blood had begun in the last decades of the nineteenth century. By the early 1920s an entire 'science' based on this had spread across Europe and the western world. Eugenics held that since certain peoples were of higher quality than others, it was naturally right to improve the overall human genetic strain by promoting higher reproduction among those from the superior race or class and, by extension, reducing reproduction by those less well favoured. At the time such thinking was advocated by prominent English novelists, including H. G. Wells, Marie Stopes (the founder of modern birth control) and two American presidents, Woodrow Wilson and Theodore Roosevelt.

Eugenics societies sprang up, often funded by wealthy American foundations, to promote (in the words of a 1911

Carnegie-supported research paper) 'the Best Practical Means for Cutting off the Defective Germ-Plasm in the Human Population'. Sterilisation and euthanasia were the most popular suggested methods.

It was a belief system and a climate tailor-made for the Nazis. It supported their spurious belief that Germans were the true descendants of a breed of Aryan (sometimes called Nordic) supermen whose destiny was once again to rule the world. In 1925 Hitler had promulgated this concept in his autobiographical Nazi manifesto, *Mein Kampf.*

> The products of human culture, the achievements in art, science and technology with which we are confronted today are almost exclusively the creative product of the Aryan. That very fact enables us to draw the not unfounded conclusion that he alone was the founder of higher humanity and was thus the very essence of what we mean by the term 'man'.
>
> What we must fight for is to safeguard the existence and reproduction of our race and our people, the sustenance of our children and the purity of our blood ...

Four years later, he followed this up in a speech to a party rally.

> If Germany were to get a million children a year and remove 700,000 to 800,000 of the weakest people, the final result might be an increase in strength.

It was a refrain taken up by the man who became the Führer's most powerful henchman. When Himmler was appointed head of the SS that same year, he told his senior officers:

Should we succeed in establishing our Nordic race again
in and around Germany ... and from this seed bed produce a race of 200 million, then the world will belong to
us. We are called, therefore, to create a basis on which the
next generation can create history.

One of the first pieces of legislation passed by Hitler was the
Law for the Prevention of Hereditarily Diseased Offspring.
This required doctors to register every case of hereditary illness among their female patients of childbearing age. Failure to
comply was punishable by substantial fines. The opening paragraphs of the new law set out both the problem (as the Nazis
saw it) and its primary cause.

Since the National Revolution [the quasi-legal putsch by
which Hitler gained the power to rule by decree], public
opinion has become increasingly preoccupied with questions of demographic policy and the continuing decline
in the birth rate.

However, it is not only the decline in population which
is a cause for serious concern but equally the increasingly
evident genetic composition of our people.

Whereas traditionally healthy families have for the most
part adopted a policy of having only one or two children,
countless numbers of inferiors and those suffering from
hereditary conditions are reproducing without restraint,
allowing their sick and disadvantaged offspring to be a
burden on the community.

The solution was – to Nazi thinking – obvious: sterilisation. A
system of 181 Genetic Health Courts was set up to order the

enforced neutering of those deemed substandard. A measure of the programme's immediate effect was the volume and outcome of appeals: in less than a year almost 4,000 people tried to overturn the decisions of the sterilisation authorities. Just 41 were successful. Five years later, by the start of the Second World War, at least 320,000 people had been forcibly sterilised under the legislation.

But if the draconian new law addressed the perceived problem of 'inferiors' polluting or weakening the nation's blood-stock, it did not define just what that blood-stock should be. In September 1935, a leading Nazi doctor called Gerhard Wagner announced in a speech that the government would soon introduce a 'law for the protection of German blood'. Within days this was codified into the Nuremberg Laws.

These introduced four official categories of human beings in the National Socialist state. People with four German grandparents were classified as 'German or kindred blood'; those who had one or two Jewish grandparents were deemed to have come from 'mixed blood' and were placed – in order of descending value – into two classes of *Mischling*; while anyone descended from three or four Jewish grandparents was irredeemably Jewish.

Only those who were formally registered as being the product of 'German or related blood' were now 'racially acceptable' and granted the status of *Reichsbürger* (citizens of the Reich). *Mischlings* were placed in the lesser category of *Staatsangehörige* (state subjects). Jews were from that point on deprived of all citizenship rights, and marriage between Aryans and non-Aryans was outlawed.

The Nazis proceeded to formalise these race classifications. A new set of official documents, *Der Ariernachweis* (the Aryan Certificate), were introduced to prove that the holder was a true

member of the Aryan Race. Those able to satisfy the requirement that their racial ancestry dating back to 1800 showed
that 'none of their paternal nor their maternal ancestors had
Jewish or coloured blood' were granted a *Grosser Ariernachweis*.
Others who could only produce seven birth or baptism certificates (covering themselves, their parents and grandparents), as
well as three marriage certificates from their parents and grandparents, were provided with a 'lesser' document, the *Kleiner
Ariernachweis*.

Two other pieces of paperwork became vital for life in the
Nazi state. An *Ahnenpass* was a certificate, drawn from church
records, which recorded the racial characteristics of a person's
ancestors: quite literally, an 'ancestors' passport'. It was often
supplemented by an *Ahnentafel* – a carefully tabulated version
of the ancestral family tree.

The Nuremberg Laws and the racial certificates that flowed
from them were the foundations of the Nazis' determination
to arrive at a 'final solution' for the extermination of the Jewish
population. But they were also the key cornerstones of the flip-
side of that policy: the programme to create a new Master Race
of pure-blooded Aryans who would rule Hitler's Thousand Year
Reich. The organisation which was to deliver that outcome was
Lebensborn, and its architect was Heinrich Himmler.

Himmler's papers contained his own explanation for establishing Lebensborn. His motive was, he claimed, benign and
caring.

> I have created the Lebensborn homes because I believe it
> is not right that an unfortunate girl who expects a child
> out of wedlock is kicked around by everybody ... by all
> these paragons of virtue, of male and female gender, who

feel entitled to condemn her and to mistreat her. I cannot think it right that she is being punished when the state does not provide the facility for help.

Every woman in these homes is addressed by her Christian name. One is Frau Maria and the other Frau Elisabeth – or whatever her name is. Within the homes nobody asks whether they are married or unmarried: we simply educate them, protect them and look after these mothers.

Even if this were true, I had to remind myself that Lebensborn homes were not open to every woman who found herself unexpectedly pregnant. Jewish women and *Mischlings* were excluded because they were seen as racially worthless.

As war loomed, the Reichsfüher's papers revealed a change in the purpose of the Lebensborn programme. No longer was it driven purely by the desire to increase pure Aryan blood-stock within the German population. By October 1939, Himmler had looked into the near future and seen a major threat to his plans for a future master race.

Every war involves a tremendous loss of the best blood. Many victories won by force of arms have inflicted a shattering defeat for a nation's vitality and blood. But the sadly necessary deaths of the best men – deplorable though they are – is not the worst of this. Far more severe is the absence of the children who were never born to the living during the war, or to the dead after it.

And so he issued a revolutionary order to the men under his command. In a proclamation marked 'secret' and issued to every member of the SS and police, the Reichsführer instructed them to fulfil their sacred duty to the Reich by fathering its next generation, whether or not they were married to the mothers.

Berlin, 28 October 1939

Beyond the limits of bourgeois laws and conventions, which are perhaps necessary in other circumstances, it can be a noble task for German women and girls of good blood to become even outside marriage, not light-heartedly but out of a deep moral seriousness, mothers of the children of soldiers going to war of whom fate alone knows whether they will return or die for Germany.

During the last war, many a soldier decided from a sense of responsibility to have no more children during the war so that his wife would not be left in need and distress after his death. You SS men need not have these anxieties; they are removed by the following regulations:

1.   Special delegates, chosen by me personally, will take over in the name of the Reichsführer-SS, the guardianship of all legitimate and illegitimate children of good blood whose fathers were killed in the war.

We will support these mothers and take over the education and material care of these children until they come of age, so that no mother and widow need suffer want.

2.   During the war, the SS will take care of all legitimate and illegitimate children born during the war and of expectant mothers in cases of need. After the war, when the fathers return, the SS will in addition grant generous material help to well-founded applications by individuals.

> SS-Men and you mothers of these children which
> Germany has hoped for, show that you are ready, through
> your faith in the Führer and for the sake of the life of our
> blood and people, to regenerate life for Germany just as
> bravely as you know how to fight and die for Germany.

The order did not merely authorise free sex; it demanded it.
Racially pure men and women were ordered to procreate,
whether or not they were married, so that the nation's stock of
'good blood' could be safeguarded. There was to be neither financial penalty for producing illegitimate children nor social stigma.

It is hard to overstate the radical nature of Himmler's decree.
Although the Nazis had been in power for six years and had
done much to undermine the country's family-based traditional
foundations, Germany was still a religiously conservative society.
Sex outside marriage was taboo and neither the public nor the
churches appeared ready to abandon their social mores.

Even representatives of the Nazi Party and the Wehrmacht
reacted badly to the Reichsführer's new population policy.
Yet Himmler stood firm. Three months after his 'procreation
order', he issued an unyielding and unrepentant statement to
his forces.

Office of the Reichsführer-SS and Chief of the
German Police

Berlin, 30 January 1940

*SS Order for the whole of the SS and Police*
You are aware of my order of 28 October 1939, in which
I reminded you of your duty if possible to become fathers
of children during the war.

This publication, which was conceived with a sense of decency and was received in the same sense, states and openly discusses actual problems. It has led to misconceptions and misunderstandings on the part of some people. I therefore consider it necessary for every one of you to know what doubts and misunderstandings have arisen and what there is to say about them.

Objection has been taken to the clear statement that illegitimate children exist, and that some unmarried and single women and girls have always become mothers of such children outside marriage and always will.

There is no point in discussing this; the best reply is the letter from the Führer's Deputy to an unmarried mother, which I enclose together with my order of 28 October 1939.

The Deputy Führer was Rudolf Hess. On Christmas Day 1939, the Nazi Party's daily paper, *Völkischer Beobachter,* had published an open letter to a notional unmarried mother in which Hess set out the new morality.

The National Socialist philosophy of life has given the family the role in the State to which it is entitled. However, in times of special national emergency special measures can be instituted, which are different from our basic principles. In wartime, which involves the death of many of our best men, every new life is of special importance to the nation. Hence, if racially unobjectionable young men going on to active service leave behind children who pass on their blood to future generations through a girl of the right age and similar healthy

> heredity … steps will be taken to preserve this valuable
> national wealth.

By calling on Hess, then more established in the hierarchy of
Hitler's regime, Himmler was doubtless giving himself some
political cover. But his own command of the SS was absolute
and his faith in its fundamental importance to the next generation unshakeable.

> The worst misunderstanding [of my original order] con-
> cerns the paragraph which reads: 'Beyond the limits of
> bourgeois laws and conventions …' According to this, as
> some people misunderstand it, SS men are encouraged to
> approach the wives of serving soldiers. However incom-
> prehensible to us such an idea may be, we must discuss it.
>
> What do those who spread or repeat such opinions
> think of German women? Even if, in a nation of 82 mil-
> lion people, some man should approach a married woman
> from dishonourable motives or human weakness, two
> parties are needed for seduction: the one who wants to
> seduce and the one who consents to being seduced.
>
> Quite apart from our own principle that one does not
> approach the wife of a comrade, we think that German
> women are probably the best guardians of their honour.
> Any other opinion should be unanimously rejected by all
> men as an insult to German women.

For all the Reichsführer's protestations of outrage, this was far
from an outright denial of the charge that he was promoting
sex outside marriage. Fears about lax morals in Nazi organisations had been growing for several years: in the summer of

1937, several thousand copies of a privately printed open letter to Goebbels, the Party's propaganda minister, had circulated throughout the country. Signed with the pseudonym 'Michael Germanicus', it pointedly referred to promiscuity throughout the National Socialist movement:

> ... the sexual excesses in country homes and Hitler Jugend [Hitler Youth] camps; the bad camp morals and the Bund Deutscher Mädel [League of German Maidens] girls made 'young mothers' ...

Neither the imprisonment of those caught in possession of the open letter nor Himmler's defence of his 'procreation order' ever completely suppressed this deep-seated public anxiety that Lebensborn homes were being used for sexual liaisons between SS officers and suitable Aryan mates. Himmler's own statements sometimes added fuel to this unfounded rumour, notably his description of the role of the SS in the process: 'We only recommended genuinely valuable, racially pure men as Zeugungshelfer [procreation helpers]'. It was easy, with hindsight, to see where the 'SS stud-farm' myth of Lebensborn began.

But rumours aside, the central role of the SS in the project was beyond doubt. Himmler's statement of January 1940 set out its parental role in the Lebensborn programme.

> The question has been raised as to why the wives of the SS and police are looked after in a special way and not treated the same as all the others. The answer is very simple: because the SS through their willingness to make sacrifices and through comradeship have raised the necessary funds, through voluntary contributions from leaders and

men, which have been paid for years to the Lebensborn organisation.

Following this statement all misunderstandings should have been cleared up. But it is up to you SS men, as at all times when ideological views have to be put across, to win the understanding of German men and women for this sacred issue so vital to our people and which is beyond the reach of all cheap jokes and mockery.

To properly understand Lebensborn, I needed to delve into the history and nature of an organisation which, more than fifty years after the end of the war, retained the power to instil fear and loathing. I had to immerse myself in the Schutzstaffel.

*'One basic principle must be the absolute rule for the SS. We must be honest, decent, loyal and comradely to members of our own blood and nobody else.'*

HEINRICH HIMMLER, SPEECH TO SS OFFICERS,

6 OCTOBER 1943

Wewelsburg Castle sits on a steep bluff above the rolling hills and dense forests of North Rhine-Westphalia. It was originally built for the medieval prince-bishops who ruled the *Landkreis* of Paderborn; it was they who created its unique triangular layout of three round towers connected by massive stone walls.

In November 1933, Heinrich Himmler was touring the region. Since taking charge of the SS, he had been searching for a suitable location both to house an ideological training school and to become its spiritual headquarters. When he saw Wewelsburg, he immediately decided to commandeer it.

The Reichsführer had grandiose plans for his acquisition. His obsession with Germany's past had convinced him that Westphalia was the heartland of the (entirely fictional) tradition of Aryan supermen. When he formally took control of the castle in September 1934, the *Völkischer Beobachter* informed its readers that a lavish ceremony had been held to mark the opening of an SS school dedicated to researching early Germanic history and mythology as the basis for 'ideological and political training'.

The Nazi daily paper did not report Himmler's real motivation: to create a stronghold that would, in his own words, be 'the centre of the world after the Final Victory'. Since the organisation responsible for delivering this triumph was the Schutzstaffel, Wewelsburg must be transformed into a fortress that served and glorified its mystical bonds of brotherhood.

∞

The SS began life as a small, rag-tag paramilitary force, set up to guard Hitler in the roughhouse era of the 1920s when armed Nazis fought street battles with their political opponents in the streets of southern Germany. When Himmler was promoted to its head in 1926, he was determined to transform the organisation. New and deliberately sinister-looking black uniforms replaced the previously favoured provincial *Stiefelhosen*; new rules were imposed, banning smoking and instituting military drill sessions.

By the time he ascended to the formal rank of Reichsführer-SS three years later, membership had risen from a few hundred to five thousand. Himmler drew up new criteria for recruitment. All applicants had to be at least 1.7 metres (5 feet 6 inches) tall and those who wished to enter its basic ranks had to sign on for four years, rising to twelve for NCOs and twenty-five for would-be officers. Despite these strict demands, tens of thousands of men applied.

But the height requirement and years of promised service were only the beginning. From the moment he assumed control of the SS, Himmler was determined to form its ranks exclusively with those of sound racial pedigree. He drew up a grading system by which specially appointed 'race experts' would assess the *Erscheinungsbild* (physical appearance) of each applicant, before sorting them into one of five categories: 'Pure Nordic' was at the top, followed, in descending order of worth, by 'Predominantly Nordic', 'Light Alpine with Dinaric, or Mediterranean Additions', 'Predominantly Eastern' and 'Mongrels of Non-European Origin'. Only those placed in the top three groups were considered for membership of the new SS: the remainder were rejected out of hand.

This was just the first hurdle. Those who passed the racial test were then subjected to a rigorous assessment of their other physical attributes. On a scale of one to nine, those who were assessed in the four top categories were deemed automatically acceptable; those whose fitness or physique condemned them to rungs seven or worse were shown the door; and the middle-ranking levels of five and six were graciously allowed to become SS men if their zeal for the Nazi cause outweighed their physical inadequacy.

There was, however, one criterion that was rigidly enforced. Every potential member, regardless of rank, had to be able to provide documentary proof of his racial ancestry. For enlisted men this pedigree had to stretch back to 1800; officers were required to provide evidence of their heritage from the mid-1700s. Just as ordinary citizens of the National Socialist state would soon be issued with certificates proclaiming them to be greater or lesser Aryans, every member of the SS carried a *Sippenbuch* – genealogical documentation attesting to their historic racial 'health'.

In writing this I have struggled to express the horrific

nature of the Nazi philosophy. It is easy to reach for words like 'obscene' or 'grotesque', but how does one get beyond clichés to truly convey the horror of such ideas? As a German woman raised in Hitler's Reich, I have always been acutely aware of where this obsession with race ended: in global war and devastation, of course, but also in the extermination camps of Auschwitz, Treblinka and Bergen-Belsen. I am conscious that mere words cannot do it justice, but also that I cannot shy away from this history to which my own past is inextricably bound.

The Reichsführer was not simply intent on creating a force of racially pure men: in his mind, the SS was to be the foundation of a new generation, the begetters of a Master Race. This plan had first been articulated by the man Himmler appointed to head up his 'Race and Settlement' organisation, RuSHA. Walther Darré, a former chicken-breeder who had returned to Germany from Argentina, wrote a manifesto entitled 'Blood and Soil'. In 1929, it was printed by the Nazi Party's own publishing house.

> From the human reservoir of the SS we shall breed a new
> nobility. We shall do it in a planned fashion and according
> to biological laws – as the noble-blooded of earlier times
> did it instinctively.

Himmler, who had also once been a chicken farmer, wholeheartedly endorsed this agricultural analogy, noting that by adopting Darré's principles it would be possible to 'attain the kind of success in the human sphere that one has had in the realm of animals and livestock'.

But the Reichsführer was also clear that this attempt to create an entire new generation must be kept within the SS as an organisation: he had no intention of allowing members of his precious brotherhood to mate outside its strictly guarded bonds. To ensure this, in 1932 he issued a ten-point Engagement and Marriage Decree to every member of the Schutzstaffel.

1. The SS is an association of German men of Nordic determination selected on special criteria.

2. In accordance with National Socialist ideology and in the realisation that the future of our Volk [people] rests upon the preservation of the race through selection and the healthy inheritance of good blood, I hereby institute the 'Marriage Certificate' for all unmarried members of the SS, effective January 1, 1932.

3. The desired aim is to create a hereditarily healthy clan of a strictly Nordic German sort.

4. The marriage certificate will be awarded or denied solely on the basis of racial health and heredity.

5. Every SS man who intends to get married must procure for this purpose the marriage certificate of the Reichsführer-SS.

6. SS members who marry despite having been denied marriage certificates will be stricken from the SS; they will be given the choice of withdrawing.

7. Working out the details of marriage petitions is the task of the 'Race Office' of the SS.

8. The Race Office of the SS is in charge of the 'Clan Book of the SS', in which the families of SS members will

be entered after being awarded the marriage certificate or after acquiescing to the petition to enter into marriage.

9. The Reichsführer-SS, the leader of the Race Office, and the specialists of this office are duty bound to secrecy on their word of honour.

10. The SS believes that, with this command, it has taken a step of great significance. Derision, scorn, and incomprehension do not move us; the future belongs to us!

To assure this future, prospective couples seeking Himmler's blessing had to complete an exhaustive questionnaire, detailing the colour of their hair, eyes, skin and physical attributes, and to which, even more bizarrely, they had to affix photographs of themselves in bathing costumes.

In case the purpose behind the process was unclear, the Reichsführer explicitly set out his reasoning in a directive to the SS about its responsibility to raise the new generation.

A marriage with few children is little more than an affair. I hope members of the SS, and especially its leaders, will set a good example. Four children is the minimum necessary for a good and healthy marriage.

Himmler also had a plan for those SS men who did not – or could not – produce offspring: the Lebensborn programme. In 1936, just nine months after establishing the secretive society, he placed it under the direct control of the SS. And he made clear that childless officers would be expected to help Lebensborn place at least some of the babies born in its homes.

> In the event of childlessness it is the duty of every SS leader
> to adopt racially valuable children, free of hereditary ill-
> nesses, and inculcate them in the spirit of our philosophy.

That sentence sent a chill through my bones.

The relationship between Lebensborn and the SS was not simply a matter of bureaucratic control. In the first Lebensborn prospectus, Himmler explicitly described the deep intertwining of his ostensibly benevolent society with the black-uniformed Schutzstaffel.

> The expenses in carrying out [Lebensborn's] tasks will
> be met in the first instance by member's subscriptions.
> Every SS leader attached to head office is honour-bound
> to become a member. Subscriptions are graded according
> to the SS leader's age, income and number of children …
>
> If he is still childless at twenty-eight, a higher subscrip-
> tion will be due. At the age of thirty-eight his second child
> should have arrived: if not, his subscription will again be
> increased.
>
> If at the appropriate ages further children have failed
> to appear, corresponding increases in the subscription will
> again become payable. Those who believe they can escape
> their obligations to the nation and the race by remaining
> single will pay subscriptions at a higher level that will
> cause them to prefer marriage to bachelorhood.

I found it hard to reconcile the compassionate image of maternity homes with the evil reputation of the SS. How could
anyone – even someone as blinkered and racially-obsessed as
Himmler – not have understood that the sinister Death's Head

regiments would engender fear and suspicion, not warmth and confidence? The answer, as it turned out, was that he didn't care. At the same time as he handed control of Lebensborn to the Order, he wrote:

> I know there are people in Germany who feel ill when they see our black tunic. We understand this: we do not expect to be liked by too many people.

The more I read, the more I understood that the SS was ultimately a clan, insular and secretive, moulded to fit Himmler's belief in a modern order of Teutonic knights, forever striving for the Holy Grail of racial purity. Under the Reichsführer's direction, Wewelsburg Castle was reconstructed to reflect his obsession. Rooms were named after characters from mythology – one was called 'Grail', another 'King Arthur', and in the crypt two special chambers were created. *Der OberGruppenführersaal*, the lower of the pair, was the location for the mystical rituals Himmler drew up for the twelve most senior SS leaders. Around a central eternal flame, beneath a swastika carved into the arched roof, the Reichsführer planned to hold ceremonies to celebrate death. The worship of death was central to the SS ideology, hence the Death's Head badge on their caps. It stemmed from Himmler's belief in a mystical *Götterdämmerung* – an apocalyptic vision of the destruction of the world in fire and water before its rebirth in a new and purified form.

The contrast was absurd. This was the organisation that was supposed to be in charge of nurturing and safeguarding new life in the Lebensborn homes. The Nazi regime had full confidence in its 'inevitable' success, however. Two years after the start of the war Hitler had publicly declared:

I do not doubt for a moment that within one hundred years or so from now all the German elite will be a product of the SS, for only the SS practices racial selection.

Elsewhere, Dr Gregor Ebner, a family doctor-turned-SS-officer and the man appointed by Himmler as Chief Medical Officer for the Lebensborn programme, estimated that: 'thanks to the Lebensborns, in thirty years' time we shall have 600 extra regiments'. I looked again at his prediction. A quick calculation revealed that a regiment was normally somewhere between 500 and 700 men. Six hundred new regiments composed entirely of children born in Lebensborn homes? Even at the lowest estimate that would mean 300,000 babies.

Could there really be hundreds of thousands of people like me – children of the Lebensborn programme living throughout Germany?

## TEN | HOPE

'Beware of what you wish for in youth,
because you will get it in middle life.'
JOHANN WOLFGANG VON GOETHE

From the moment the German Red Cross asked if I was interested in finding out about my family, I had thought of little else. But as the months crawled past with no response to the letters I had written, I came to accept that I had been searching for much longer. Looking back, my whole life seemed to have been overshadowed by secrecy. No matter how hard I worked, no matter how much I gave of myself to the poor, damaged children who came to my practice, nothing could free me from the unhappiness of not knowing who I was. And so I had longed and hoped and dreamed.

The phone call from the Red Cross had broken the spell. No longer was I half-asleep, seeing snatches of my past only in dreams: the promise of solid, reliable information had woken me. And I yearned all the harder for it.

'Be careful what you wish for', warned Germany's most famous poet, writer and statesman. Perhaps I should have listened to Goethe.

The letter arrived in October 2000. It was from a Jože
Goličnik, the director of an archive in Maribor, Slovenia's second city and the capital of the Lower Styria region. I had heard
that an old repository of parish documents was held there and,
having heard nothing back from the government of Slovenia,
had decided to try my luck with the church. I'd written purely
on chance, with little hope that it would yield anything useful.
But Mr Goličnik said he had found a record of my family.

> The father of Erika Matko is Johann Matko from Zagorje
> ob Savi. Her mother came from Croatia. Mr Johann
> Matko lived in Sauerbrunn and was a glass-maker.

Sauerbrunn. It existed. Not in Austria but in Slovenia – or,
more accurately, in the old Yugoslavia. I was so happy that,
quite spontaneously (and most unlike me), I literally burst into
song, full of relief and excitement. Of course I still had to locate
Sauerbrunn, which was not likely to be called that now. The fall
of communism had been slower in Yugoslavia than elsewhere
across the eastern bloc, but when it happened it brought civil
war in its wake. As the smoke cleared from the bloody years in
which Serbs fought Croats, Bosnians, Montenegrins and all the
other nationalities Tito had welded into a unified republic in
the 1940s, new nations rose from the ashes: many changed the
names of their towns and cities.

Mr Goličnik's letter had contained a clue, though. Johann
Matko had been a glass-maker. If I could find a region that
had contained major glass factories, I stood a chance of tracking down Sauerbrunn and finding its new name. Better still,
attached to the letter was an actual copy of the parish records,
including the Matkos' birth dates. Johann had been born on

12 December 1904; his wife – Helena Haloschan – was eleven years younger, born in St Peter, Croatia, on 8 August 1915.

I wrote again to the government of Slovenia, updating my original request with the information from Maribor. But I decided my best bet was to contact the German Red Cross. They had effectively instigated my investigation and I felt hopeful that if anyone could help me track down the Matkos and Sauerbrunn, it would be their staff. I sent off the latest in my growing volume of letters. And then I began to research glassmaking in the former Yugoslavia.

It didn't take me long to discover that glass had been a regional specialty of Lower Styria for more than three centuries. From the 1700s onwards, factories had sprung up producing highly prized and beautifully crafted lead crystal. The centre of this tradition was the town of Rogaška Slatina. And its previous name? Sauerbrunn. I had tracked down the birthplace of Erika Matko; I had found my home.

I cannot describe the elation I felt at that moment. At long last it seemed as though I could almost reach out and touch my biological parents – surely soon I would be able to do so in real life.

Be careful, said Goethe.

My optimism lasted only a few weeks before reality intruded. The Red Cross replied to say they had no information about anyone called Matko from Yugoslavia in its records of those whom the Nazis had captured or killed. The letter advised that the organisation was unable to carry out any research in the archives of former Yugoslavian countries and warned that if I chose to do my own research, there was a very high probability I would discover that Erika Matko's parents were dead, and that they had not died of natural causes. The message was clear:

whatever records of their existence might once have existed, my parents were likely to have been killed by the Nazis after Hitler's armies invaded Yugoslavia and, most likely, any trace of them would have disappeared at that point.

The note from the Red Cross was not the worst of it. In February 2001 I received a letter that dashed all my hopes of ever finding my family. The Slovenian government had been neither quick nor helpful in the months since I had first written asking for information. Now, when it did finally send a substantive response, the words hit me like a blow to the stomach.

> We wish to inform you that, according to the local admin-
> istration of Rogaška Slatina, they have discovered [records
> of] an Erika Matko, born on November 11, 1941. But
> this woman is still living inside Slovenia: therefore the
> assumption that Ingrid von Oelhafen was born as Erika
> Matko is wrong.

The generation from which my parents (whoever they were) came had known more physical suffering than I would ever experience, but still it seemed to me that the cruellest pain of all was that of being offered hope, tentatively reaching out to grasp it – and then seeing it snatched away.

I sat at the table in my flat, the letter in my hand, as my dreams dissolved in front of me. Please believe me: this was not merely self-pity. I had known for decades that I was not really Ingrid von Oelhafen. I had salved that wound with the knowledge, derived from the few scraps of paper that had travelled with me through the years, that I was once Erika Matko. Now I was neither Ingrid nor Erika. I was, truly, no one.

When the shock wore off I took time to reflect on the toll

this quest was taking. I forced myself to look at how the ups and downs of my investigation were affecting me and realised that I had spent a whole year riding an emotional roller coaster, soaring high one minute, plunging down the next. Was it really worth the pain? I had made a life – a successful and generally happy life – as Ingrid von Oelhafen, regardless of my true origins, and I had official German papers which said that I was Ingrid. Really, what did it ultimately matter that I might – or might not – have once been called Erika Matko? Would it make me happier to continue pursuing the mystery of the Matko family and the country where this Erika had started life?

I decided that the answer was no. I bundled my letters and research notes into a file and put it away in a drawer. I resolved to forget about them, at least for the time being. Even when the archivists at Bad Arolsen later wrote to me with the news that they had, after all, found some documents relating to Erika Matko and Lebensborn, I simply filed the letter away with the other documents.

Months flew by, then a full year. I buried myself in work and studying music. I had been learning to play the flute; now I practised harder, immersing myself in the notes and melodies of the classical composers.

By the time another envelope dropped onto my doormat, a year and a half had passed since I had put away the folder marked 'Erika Matko'. Perhaps if the letter had been from anyone else it might have joined the others. But this note was from Georg Lilienthal and it contained an invitation. For the first time ever, a group of Lebensborn children was to meet: would I like to come?

Would I? Honestly, I was not sure. My journey so far had been full of dead ends, false trails and seemingly insurmountable obstructions. Did I really want to risk opening up old wounds all over again? And even if the answer was yes, what did I actually have to contribute? I took out my abandoned file of paper, with its obscure Nazi documents and contradictory modern correspondence: what could I really tell anyone? I agonised about it, turning the whole business over and over in my mind.

In the end, I realised that I had no choice. The questions about where and who I came from had been part of my life ever since the day Frau Harte had told me I was not Hermann and Gisela's real daughter. They had been at the back of my mind, pressing down on my emotions – and perhaps shaping my actions – for more than fifty years. Hiding from them simply didn't work: I would have to go to the meeting.

In October 2002, I packed up my car and began the long drive south: it was 260 kilometres from my home in Osnabrück to the town where the meeting was being held. I was sixty-one years old, and it was time to learn about my childhood.

_'We aren't perfect. We've got all the same
illnesses and disabilities as other people.'_
RUTHILD GORGASS, LEBENSBORN CHILD

Between Cologne and Frankfurt, the small town of
Hadamar sits on the southern edge of the Westerwald,
the long, low mountain range running down the eastern bank
of the river Rhine.

It is known today for its highly regarded institutes devoted
to forensic and social psychiatry, and for a stark obelisk com-
memorating the victims of the Nazis' Aktion T-4 euthanasia
programme, which had been based in the town. Through their
research, Georg Lilienthal and other historians had revealed
that between 1941 and 1945, thousands of disabled or otherwise 'undesirable' men, women and children were brought to
Hadamar to be sterilised or put to death.

Although Aktion T-4 officially ended in 1941, the programme had, in fact, continued until the Nazis' surrender
in 1945. In total, nearly 15,000 German citizens were sent to
Hadamar's hospital: most were subsequently murdered in a gas
chamber. This was the town where I was to meet the other
Lebensborn children.

They were not, of course, children any longer. Like me, the twenty men and women sitting around the room that morning in October were in their sixties and close to retirement. As I took my place, I was very nervous. One by one, we introduced ourselves: when it came to my turn I made myself speak the single sentence I had rehearsed. 'My name is Ingrid von Oelhafen. I don't know anything.' And then I burst into tears.

My new companions were kind and caring. Each was much further into their personal investigations than I, and because they had been through the same emotions they understood my anxiety. As they told their stories, the callous brutality of the Lebensborn programme became clearer to me; and though each new revelation was shocking, learning the truth also somehow put me at ease.

Ruthild Gorgass had been one of the first Lebensborn children to look for others who had been born or brought up in the programme. She was around my age; tall with blue eyes and a brush of short blond hair. She was a physiotherapist too and, like me, she had inherited a diary kept by her mother, which had helped her understand the story of her birth. I liked her immediately and felt comforted by her presence.

Her story was also a good introduction to Lebensborn. Ruthild's father was forty-nine when she was born. He had been a lieutenant in the German army during the First World War. In 1916 he was badly injured at the battle of Verdun, his back and chest a mass of shrapnel splinters.

In the 1930s he had become a committed Nazi and by the start of the Second World War he was a big shot in the chemical industry. He was also married with a teenage son. Despite this, at some point he met and began an affair with Ruthild's mother, who was a clerk in the Leipzig Chamber of Commerce, eighteen years younger than him. Just before Christmas 1941, she found that she was pregnant. Her position fitted precisely Himmler's original aim for Lebensborn: both her parents were dead, she was unmarried and carrying an illegitimate child and therefore at risk of the opprobrium of her family and prejudice from her community. Above all, her child's father was a card-carrying Nazi, and both he and Ruthild's mother were able to demonstrate their genealogical racial purity. In the summer of 1942, the two of them made the 170-kilometre journey from Leipzig to Wernigerode, a small town deep in the spectacular Harz Mountains of Saxony. There, in the heartland of old Germany, Himmler had established a Lebensborn maternity unit. In August 1942, Ruthild was born.

Heim Harz, I learned, was one of twenty-five Lebensborn homes established across Germany and in the countries its armies overran. There were nine homes in Germany itself, two in Austria, eleven in Norway, and one each in Belgium, Luxembourg and France. Often they occupied buildings taken from Hitler's political enemies or wealthy Jewish families: the organisation's central headquarters in Munich had belonged to the writer and exiled anti-Nazi activist, Thomas Mann. Some of the premises were furnished with property confiscated from people who had been sent to the death camps, and each was equipped with state-of-the-art medical equipment to ensure that Himmler's precious pure-blood babies were delivered safely into the world.

And they were. In 1939, Dr Gregor Ebner, Lebensborn's chief medical officer, sent a report to Himmler detailing the success of the programme. More than 1,300 pregnant women had applied to give birth in the homes. Racial and hereditary health examinations had reduced this number by half, so that a total of 653 mothers-to-be were admitted. The neo-natal mortality rate for Germany as a whole at the time was 6 per cent: in the Lebensborn homes this figure was cut in half.

> The births are very easy, without many complications.
> This is attributable to the racial selection and the quality
> of the women we get.

Their success came at a price, however. Ebner reported that the cost per mother was a substantial 400 Reichsmarks. But, he noted, 'that isn't much of a sacrifice if you can save a thousand children of good blood'.

Blood was all-important. Lebensborn was charged with ensuring a racially selected future master race to rule over the global empire of Hitler's Thousand Year Reich. There was even a slogan that encapsulated the duty of the women who gave birth in the homes: '*Schenkt dem Führer Ein Kind*' (give a child to the Führer).

The physical health of the Lebensborn mothers may have been uppermost in Himmler's mind, but he was also determined to monitor and guide their political wellbeing. To ensure that they left the homes even more zealous than when they arrived, women were required to attend three sessions of ideological 'education' every week of their stay. During these classes they watched propaganda films, read chapters from *Mein Kampf*, listened to radio lectures and took part in communal singing of party anthems.

Staff members were themselves carefully monitored and instructed to complete detailed questionnaires about each of the mothers under their care. These RF-Fragebogen (the initials stood for Reichsführer) recorded every aspect of the women's

personalities, from their behaviour in the home to their brav-
ery (or otherwise) during birth and their commitment to the
National Socialist cause: each document was sent to Berlin,
marked for the personal attention of the Reichsführer-SS.

This was not just a bureaucratic nicety. Even in the midst
of the war – at a time when he was overseeing wholesale murder in the death camps and the entire apparatus of the Nazi
terror throughout Europe – Himmler applied himself devot-
edly to these questionnaires, deciding, on a case-by-case basis,
whether a woman would be allowed to give birth to a second
child in Lebensborn at any point in the future. In fact, he
supervised every aspect of life in the homes, from the trivial to
the absurd. On one occasion he instructed his personal aide,
SS-Stamdartenführer Rudolph Brandt, to write to the head of
Lebensborn demanding that a record be kept of nose shapes.

> The Reichsführer-SS wants a special card index to be
> kept of all mothers and parents having a Greek nose or
> the rudiments of one. As an example of the type required,
> you should refer to the mother in Questionnaire L6008,
> Frau I.A.

Himmler's hands-on control extended to diet. He issued a
stream of memos, instructing cooks on the correct way to steam
vegetables and demanding that the homes' supervisors make the
women eat porridge – apparently because he had identified this
as a vital factor in forming the racially admirable characteristics of the English aristocracy. For good measure he insisted
on the application of regular doses of cod liver oil, much to
the evident disgust of the recipients. He regularly visited the
homes, checking upon the progress of the women and their

children. So complete was his involvement that babies born on Himmler's birthday were formally registered as his godchildren and received a special memento – a silver cup, engraved with his name as well as that of the baby.

I found these bizarre details of life in Lebensborn homes bewildering. How did the second most powerful man in the Reich find the time to control day-to-day life in twenty-five maternity homes?

But beyond the oddities, the stories told by the Lebensborn children that day in Hadamar revealed the programme's darker elements. Ruthild told me that she and other children underwent a quasi-religious naming ceremony in which they were dedicated to Hitler and the brotherhood of the SS. This *Namensgebung* ritual was a distorted version of the traditional Christian baptism, with an altar draped in a swastika flag and a bust or photo of the Führer in pride of place. In front of a congregation made up of Lebensborn staff and black-uniformed SS officers, mothers like Ruthild's promised that their children would be raised as good National Socialists: they then handed over their babies to an SS man who intoned a 'blessing'. There appeared to be different versions of this liturgy in different homes, but the essence of each was the same.

> We believe in the God of all things
> And in the mission of our German blood
> Which grows ever young from German soil.
> We believe in the race, carrier of the blood,
> And in the Führer, chosen for us by God.

An SS dagger was held over the baby and the senior officer read out a formal welcome to the brotherhood of the SS.

We take you into our community as a limb of our body.
You shall grow up in our protection and bring honour to
your name, pride to your brotherhood and inextinguish-
able glory to your race.

How could a mother hand over her precious baby to the
care – if that's what it was – of an organisation like the SS?
What parent could do something so horrific? As I had told the
gathering right at the start, the only thing I knew about my
origins was that I had been a baby in the Lebensborn home
at Kohren-Sahlis: had I too been dedicated to the service of
the Nazis?

There were more terrible revelations to come. I already
knew what Himmler was planning with the Lebensborn programme. But I had not realised just how far his organisation
would go to ensure that the new *Herrenrasse* – this master race
– was free of any physical defect.

They called them *Kinderfachabteilung*. Literally translated, this
means 'children's ward'. It sounds such an innocent phrase, but
it wasn't. Under the Aktion T-4 euthanasia programme, babies
born in the Lebensborn homes with developmental delay, disease or mental disabilities were killed.

Jürgen Weise was born in the Lebensborn home at Bad
Polzin on 5 June 1941. The head of Lebensborn – a Nazi
named Max Sollman – ordered that Jürgen be taken to a
*Kinderfachabteilung* in Brandenburg, near Berlin. There he was
given tranquillisers and deliberately left untended and unfed. On
6 February 1942, the little boy died; he was eight months old.

Jürgen Weise was not the only disabled baby to be murdered
in the name of racial purity and strength. In 2002, when we
met, research into this was at an early stage, hampered by the

reluctance of staff at official archives to allow access to Nazi-era documents.

But the Brandenburg *Kinderfachabteilung* had been exposed several years before, and there was convincing evidence that 147 babies were murdered there – including an unknown number from the Lebensborn homes.

I struggled to take all this in. I had dedicated my life to disabled children. I had seen the joy that my efforts brought to them and to their parents. I had felt the love that comes from helping children like Jürgen. What sort of heartless bureaucrat could so easily extinguish such precious life?

Perhaps my reaction sounds naive. History has told us that the Nazis ruthlessly and quite openly murdered millions of Jews in Auschwitz and Bergen-Belsen and the other camps: why would the deaths of a few babies, born in secret and hidden from public view, matter to men like Himmler and Hitler? But according to their own twisted ideology these children were special: they were in the Lebensborn homes because their parents had been examined and ultimately proven to be suitable blood-stock for the master race.

I looked around at the men and women who had also been part of the Lebensborn programme. I assessed each body, searched each face for evidence that these survivors of Himmler's experiment were truly super-human beings. Were they taller, stronger, healthier than anyone else?

Ruthild answered my unspoken question. She took off her glasses, rubbed her eyes and said: 'We aren't perfect. We've got all the same illnesses and disabilities as other people.'

What, then, was it all for? Himmler's great dream was a generation of super-Aryans who would be so strong, so flawless that they would grow up to become the natural aristocracy of the National Socialist state and the lesser nations it ruled. Yet his scheme seemed to have produced nothing more notable than the group of perfectly ordinary men and women seated around me.

There were, though, two striking characteristics shared by most of these Lebensborn children: deep emotional hurt and a palpable sense of shame. The former stemmed from a problem I was very familiar with. Each of us who began life in that clandestine programme had grown up with the pain of not knowing the truth about our birth.

This secrecy was both deliberate and carefully managed from the outset. Doctors and staff in the Lebensborn homes were required to swear an oath of silence which committed them to respecting 'the honour of pregnant women, whether they conceived before or after marriage', and in June 1939, Himmler had issued an order to protect the identity of illegitimate children born in the programme.

> Following an agreement between the genealogical office Reich Minister of the Interior and the L organisation, it is possible to maintain secrecy about the origin of illegitimate children born in Lebensborn homes for an unlimited period. The Reich office will provide a certificate confirming the child's Aryan descent. This certificate can be produced by a child born in a Lebensborn home when they start school, for the Hitler Youth and for institutions of higher education, without the slightest difficulty arising.

This determination to throw a cloak of confidentiality over all aspects of the Lebensborn children extended to the records kept of their delivery. Himmler's organisation set up a special registry office for recording the births: this was kept separate from any other office of the Reich and operated in total secrecy. Mothers' names might be shown in the files, but the identity of the father was generally deliberately omitted.

And many of these deliberately redacted records had themselves disappeared: in the final days of the war, with Allied forces closing in on the homes, Lebensborn staff destroyed much of the organisation's paperwork. As a result, most of the children born within Lebensborn grew up not knowing who their fathers were – and, unless their mothers broke the bonds of secrecy, completely unable to find out.

In particular, this affected the children who had been handed over to foster parents. But even those who, like Ruthild, were kept by their biological mothers, often found it impossible to prise out the information. Many mothers were very vague about their time in Lebensborn homes; others refused point-blank to discuss it.

I knew how that felt. Although I didn't yet fully understand how I fitted into the story of Lebensborn, I was familiar with parental walls of secrecy: as Georg Lilienthal had warned me, Gisela undoubtedly withheld much of what she knew throughout my life.

Why did other mothers do this too? The reason was the second characteristic evident in many of the Lebensborn children sitting beside me in Hadamar. Shame is a powerful emotion, and the political climate in post-war Germany was hardly conducive to honesty about involvement with an organisation as reviled and feared as the SS.

One of the men in our group talked openly about the guilt and the shame that had blighted his life. His story opened my eyes to another aspect of the Lebensborn programme. Hannes Dollinger had grown up in Bavaria, where the couple he thought of as his parents owned an inn. But after he started school, he heard rumours that he was a foundling. He asked his parents whether this was true, but they refused to answer and when he persevered they punished him and forbade him from ever raising the subject again.

It was not until he was fifty that he learned the truth. Just as Frau Harte had once broken the news to me that Hermann and Gisela were not my real parents, a former employee of Hannes' family told him on her deathbed that he had been adopted. That alone was a shock, but the story of how he came to Bavaria was worse.

Norway was the northernmost country occupied by Hitler's army. The Wehrmacht invaded in April 1940 and from then until the end of the war, Norway was run by a collaborationist government which enthusiastically did the Nazis' bidding.

Himmler had for many years viewed the largely blond and blue-eyed local population as de facto Aryans. He and his officials actively encouraged liaisons between SS or Wehrmacht troops and Norwegian women, establishing a network of Norwegian Lebensborn homes in which the resulting babies were born, then shipped back to the Reich and handed over to suitable couples either for adoption or fostering.

The legacy of this collaboration was long and bitter. Unlike the desperate bonfires built by Lebensborn staff across Germany, in Norway the SS never managed to destroy its files. As a result, after the war thousands of Lebensborn babies and their mothers were identified and faced the fury of their countrymen. Women and their children were harassed by their neighbours or schoolmates. Police arrested between 3,000 and 5,000 women who had slept with German soldiers and marched them off to internment camps. The head of Norway's largest mental hospital publicly stated that women who had mated with Germans were 'mental defectives' and declared that 80 per cent of their children were retarded.

Hannes discovered that he was one of these children. He began researching his origins and found that his real name was Otto Ackermann and that he was born in September 1942 in a Lebensborn home near Oslo. From there he had been sent like a parcel across Germany, first to a Lebensborn home in Klosterheide, near Berlin, then on to Kohren-Sahlis, the home in which I had been raised.

Eventually, after being transported to the Lebensborn home in what was now Poland, he was handed over to his adoptive parents in Bavaria. It took him many years to retrace this long and complicated route. Finally he managed to obtain the name of his biological mother, but by the time he discovered this, she had died. His father, a Wehrmacht soldier, had been killed in the last months of the war and his adoptive parents had also passed away.

In many ways, Hannes was typical of our generation of Germans – ironically, since by birth he wasn't actually German. He was then a local government official and a stickler for doing things by the book. He informed the federal government of his real name and that he was originally Norwegian, and asked to change his identity documents to make them accurate. For his trouble, the government declared him stateless – and by law stateless people were forbidden from employment in any public office. It took two long and difficult years before he was offered German citizenship. Even then, to get it he had to give up his original real name.

It was heartbreaking to listen to Hannes' story. So much of it mirrored my own life – the home at Kohren-Sahlis, the problem of being declared stateless – but his experiences seemed to have been much worse. I began to feel almost lucky, and perhaps grateful, that I knew so little about where I had come from.

But at the same time the question was still hanging over me. I had learned a great deal about the Lebensborn programme and about life in its homes, but I did not know how I fitted in to this history. The documents I had found in Gisela's room showed that I had been fostered as part of something called 'Germanisation'. Neither Ruthild's story nor Hannes' contained anything to shed light on this mysterious word.

And then another member of the group stood up to speak, and I began to see the worst horror of Himmler's terrible experiment – and how I had come to be a part of it.

Folker Heinecke was six months older than me. He was a big, well-dressed man who had made a small his fortune as a shipping broker in Hamburg and London. Though he was financially well off, for much of his life he had been deeply troubled by the knowledge that he had been raised in a Lebensborn home, before being put up for adoption by the home in 1943 before the age of three.

> My first memory is of being in a room with thirty other children. I remember these people coming in, while we were lined up like pet dogs to be chosen for a new home. The people were to be my parents. They went away and came back a day later. My 'mother' apparently wanted a girl, but my 'father' wanted a boy who would be able to take on his family business in the future. I laid my head on his knee and that did it for him – I was to be their son.

Folker's new family was financially secure and well connected. Adalbert and Minna Heinecke were fanatical Nazis and owned a successful Hamburg shipping company. Adalbert was also deaf and, under Lebensborn's strict rules, should not have been allowed to foster – let alone adopt – one of its precious children.

But Adalbert was also a personal friend of Heinrich Himmler: both the Reichsführer and Martin Bormann (Hitler's personal secretary and one of the most powerful men in the Nazi regime) visited the family's home.

Like many other Germans, the Heineckes kept a small flock of hens. As Himmler had once been a chicken farmer and was a firm believer in applying the principles of poultry breeding to the human species, it was only natural that he and Adalbert talked as they studied the family's birds. When they had finished, Himmler agreed to rubber-stamp Folker's adoption.

Folker remembered a happy childhood, insulated from hunger or want due to his family's wealth. Even at the height of the Allied assault on Germany, when he watched RAF bombers weave through the flak and searchlights to launch raids into enemy territory, his chief memory was of finding the war exciting. It was not until after the war that he discovered he had been adopted.

> One of the local kids I was playing with said: 'You know you're a bastard, don't you, they're not your real mum and dad.' But back then I didn't really know what that meant.

His parents never talked to Folker about where he had come from or how he came to be adopted. When his father retired, he took over the family shipping business and enjoyed a successful career.

In 1975, after his parents died, he found among his father's papers a series of official documents he had never seen before. These recorded that he had been born at Oderberg in Upper Silesia: this area had been annexed into Hitler's Reich, but after the war it had been transferred back into the territories of the new republic of Poland. The papers also indicated (falsely, as it turned out) that both of his biological parents had died – hence the need for his adoption.

The discovery prompted Folker to investigate his origins. He approached the German Red Cross, the British Army of Occupation, the American authorities and more than thirty other agencies and church offices: slowly he began piecing together the confusing jigsaw of his past. But Poland was then still locked away behind the Iron Curtain, and it was difficult even for a man of his wealth to gain access to its archives. It wasn't until the fall of communism and the restructuring of Eastern Europe after 1989 that he finally unearthed the truth.

By 1941, Himmler's great hope that the Lebensborn programme would produce tens of thousands of racially pure babies was fading fast. In part this was due to the rigorous selection criteria, which led more than half the pregnant women who applied to give birth in the network of homes to be rejected.

Nor had the SS lived up to its leader's expectations: far from meeting the minimum requirement of fathering at least four children, the birth rate stayed stubbornly around an average of 1.5 per man. The 600 new battalions of Lebensborn babies predicted by its chief medical officer Gregor Ebner were a long way off – if they were achievable at all.

The Thousand Year Reich needed its future warriors to survive. Hitler's vision had always been for a total and global war, followed by permanent occupation of conquered lands.

But by 1941 the war was claiming thousands of German lives a week. The Lebensborn homes could not hope to fill the gap. And so Himmler decided on a new strategy: he issued secret instructions to his troops and officials to kidnap 'racially valuable' children from the countries they overran.

The wholesale stealing of children – could it really be true? Shockingly, it was: there was even a recording of Himmler giving a private speech to SS officers in which he justified the policy.

> What good blood there is of our kind in these peoples, we will take in; we will steal the children if necessary and bring them up here with us.

The name given to this plan was Germanisation. It was the word on my documents, which I had never understood: now I began to learn what it meant in practice.

Folker's tragedy was that at the age of two he looked German: he had blond hair, blue eyes and looked for all the world like a pure-blooded Aryan boy. Because of this he was snatched from his parents by SS officers and taken to a medical institute for full racial assessment.

> I was measured everywhere – head size, body size – and doctors checked to be sure that I had no 'Jewish Aspects'. When I passed those tests the Nazis declared I was capable of being Germanised and shipped me off to a Lebensborn home in the Fatherland.

After a brief stay in Bad Polzin, he was then sent on to Kohren-Sahlis. There were no completely reliable dates for his arrival there, but from what he had discovered it sounded as

though we must have been in the home at the same time. The
possibility excited me and I tried hard to unearth any memories from the back of my mind, but I could recall nothing
more about the place. It was so frustrating to meet someone
with whom I might have shared my earliest years and yet to be
unable to dredge up any recollection of the place or people.

There were other similarities in our stories. Folker's documents indicated that he had been picked up from Kohren-Sahlis
by the Heinecke family. That was where Hermann and Gisela
had come to collect me. Had I been lined up for inspection,
one of the 'pet dogs', as Folker put it, to be examined by my
prospective foster parents?

Folker's investigations had also revealed that Lebensborn
frequently gave new identities to the kidnapped foreign babies
shipped to its homes and issued false documents proclaiming
them to be either German orphans or ethnically Aryan children
of the German diaspora: *Volksdeutsche*. Once again I recognised
the word Folker used: it was on the papers I found when clearing out Gisela's room.

The clues were beginning to add up. According to my documents I had been born Erika Matko, a *Volksdeutsches Mädchen*.
I had been brought from Sauerbrunn to the Lebensborn home
at Kohren-Sahlis for Germanisation, before being handed to
the von Oelhafens to be raised as a 'real' German girl. I was
part of the scheme to '*Schenkt dem Führer ein Kind*'. I was one
of Hitler's children.

It was horrific – and yet for the first time in years I felt
I was finally getting close to solving the mystery of where I
came from.

The biggest question I had been struggling with was whether
I really was (or had once been) Erika Matko. The Slovenian

government's response to my letter had seemed to prove that I could not be, and yet the few original documents I had been able to find all showed the opposite.

Folker Heinecke's story suggested an answer to the puzzle. His investigations had led him to believe that the name recorded on his Lebensborn papers might not be genuine and that the birthplace listed might also be false. The Lebensborn head office evidently went to great lengths to erase the original identity of those children stolen from the Reich's occupied territories. Folker had discovered documents in the archives of the Nuremberg War Crimes Tribunal that told the story of a baby named Aleksander Litau, stolen from parents living in Alnova on the Crimean Peninsula. There were strong indications that Folker might have been this child: the dates matched, as did the Lebensborn homes the little boy had been shipped off to. Could I have suffered a similar fate?

Folker had ultimately run into the same bureaucratic stone wall that had previously defeated me: the documents he needed to confirm or disprove this theory were almost certainly somewhere in the ITS files in Bad Arolsen, but the archive was still not yet fully open. It was, he told us, painful to be so close and yet still so far away from the truth.

> All I really want is to find the grave of my biological mother and father. I don't want to end up driven bitter and mad by wondering what might have happened to them. I just want to know who I was and what I might have been if things hadn't turned out the way they did.
>
> I have to keep searching to find something that might lead me to who my parents really were and where they

are buried. Then I will have done my duty as a son. I will
have honoured my real parents.

I was determined that I too would keep searching and that
one day I would track down my true family. Meeting the
other Lebensborn children – fellow survivors of such a terrible experiment – gave me renewed strength to restart my
own investigations. Now I knew how and where to begin:
Nuremberg.

There was one final conversation I needed to have before
making the long drive home to Osnabrück. One of the few
non-Lebensborn people in the room that day was a man called
Josef Focks. I had not heard of him before but he had established
a reputation for tracking down documents and information
about families who had become separated during or after the
war. For his efforts the press had nicknamed him 'the Father
Finder'.

I spoke with him briefly and explained my situation. I
described the difficulty I had faced in getting information
from official archives and told him about the documents I
possessed showing me to have once been Erika Matko from
St Sauerbrunn, and how this had been explicitly contradicted
first by the Austrian authorities and then by the government
of Slovenia.

Herr Focks listened and took notes. When I finished, he
agreed to help me.

I was grateful, of course, but to be completely honest I was
thinking more about the enquiries I would make in Nuremberg
than what the Father Finder might unearth.

# TWELVE

# NUREMBERG

'Lebensborn was responsible, amongst other
things, for the kidnapping of foreign children
for the purpose of Germanisation ... numerous
Czech, Polish, Yugoslav and Norwegian
children were taken from their parents.'

INDICTMENT: THE NUREMBERG
MILITARY TRIBUNALS, CASE 8

It was spring 2003 by the time I set off for Nuremberg, 500
kilometres south.

Nuremberg was the dark heart of National Socialism.
Between 1927 and 1938 it was the city where Hitler held spectacular torchlit rallies – serried ranks of tens of thousands of
supporters screaming 'Sieg Heil' beneath an ocean of swastika banners, all captured in melodramatic propaganda films
– and where the 1935 Race Laws that signalled the start of the
Holocaust were promulgated.

For the Nazi Party, Nuremberg's position at the centre of
the country symbolised, in some mystical way, the connection
between the Third Reich and the supposed Aryan supermen
of Himmler's imagination. It was also heavily fortified, which
made it one of the last cities to fall to the Allied forces in

the final weeks of the war. Despite systematic bombing that destroyed 90 per cent of the medieval centre, the city was only captured after four days of fierce house-to-house fighting.

The three main Allied Powers, the Soviet Union, the United Kingdom and the United States, had long planned to mount public trials of the Nazi leaders, even before the war ended. On 1 November 1943 they published a joint 'Declaration on German Atrocities in Occupied Europe', issuing 'full warning' that as and when the Nazis were defeated, the Allies would 'pursue them to the uttermost ends of the earth ... in order that justice may be done'.

For the next eighteen months, as their armies slowly inched to victory, lawyers and politicians from all three countries hammered out a set of innovative legal principles under which leading Nazis could be prosecuted for war crimes and crimes against humanity. When the war ended, the only remaining issue was where to hold the hearings.

Leipzig and Luxembourg were briefly considered and rejected. The Soviet Union favoured Berlin – the 'capital of the fascist conspirators' – as a suitably symbolic location, but the overwhelming destruction suffered by the city made it impractical. The decision to choose Nuremberg was based on two key factors. Its role in the Nazi propaganda machine made it an appropriate site to dispense exemplary justice but, more importantly, its spacious Palace of Justice had survived the war largely intact – and its buildings included a large prison facility.

The surviving leaders of the Third Reich were brought to the cells beneath the courtroom in November 1945. Hitler himself had cheated justice, committing suicide in the Führerbunker amid the flames and ruins of Berlin. Himmler, too, had taken the coward's way out, swallowing a cyanide capsule while in captivity. But twenty-two others, including Reichsmarschall Hermann Göring and Deputy Führer Rudolf Hess, were brought before the International Military Tribunal and arraigned for the crimes of the Nazi regime. Eleven months later the judges, one each from France, Britain, America and the Soviet Union, pronounced the verdicts. Twelve of the accused were sentenced to death; seven received prison sentences ranging from ten years to life in prison; and three were acquitted (two trials did not proceed, one defendant having killed himself and another declared unfit for trial). On 16 October 1946, the executions were carried out in a gymnasium attached to the court building.

It was to this forbidding complex that I was heading on a spring morning in 2003. The reason for my journey was not the famous trial itself, but a much less celebrated set of proceedings held in the same building.

Although the Allies had initially planned to hold a lengthy series of joint-power tribunals, the looming Cold War and the freeze in relations between East and West made this impossible. While the main trial was still in progress, the United States took a unilateral decision to mount subsequent hearings for so-called 'second-tier Nazis' on its own. The result was a series of twelve separate prosecutions, between 1946 and 1949, in which a total of 183 defendants were indicted. Among them were the leaders of Lebensborn.

Within days of returning home from Hadamar, I wrote to the office that kept records of the Nuremberg tribunals. Given my previous experience of German official archives, I was not expecting much in the way of a response, and so I was pleasantly surprised to receive a letter from the archivists which said that there was an entire box full of relevant documents and asked whether I would like to examine them.

The papers turned out to be from Case 8 of the subsequent prosecutions. This was formally titled *The United States of America vs. Ulrich Greifelt, et al*, but was more usually referred to as the RuSHA trial, since all fourteen defendants had held senior positions in the Rasse-und-Siedlungshauptamt [RuSHA]: the Race and Settlement Main Office Himmler had established to safeguard the 'racial purity' of the SS and which had then gone on to administer Lebensborn.

I began by examining the official indictment, served on 10 March 1948. It was long and detailed – fourteen closely typed foolscap pages setting out charges under three separate headings: crimes against humanity, war crimes and membership of the SS, which had been declared a criminal organisation. It started with a bleak, uncompromising assertion:

> Between September 1939 and April 1945, all the defendants herein committed Crimes Against Humanity ... The object of this programme was to strengthen the German nation and the so-called 'Aryan' race ...
>
> The SS Main Race and Settlement Office (RuSHA) was responsible, amongst other things, for racial examinations. These racial examinations were carried out by RuS leaders ... or their staff members called racial examiners (Eignungsprüfer) in connection with ... [the] kidnapping of children eligible for Germanisation ... Lebensborn was responsible, among other things, for the kidnapping of foreign children for the purpose of Germanisation.

It was chilling to read these words. Here, in the cold legal language of a trial, was the essence of the organisation that had cared for me – if that was the right word – in my earliest years.

The indictment went on to set out both the motivation for the programme, and the countries in which it had operated.

> An extensive plan of kidnapping 'racially valuable' alien children was instituted. This plan had the two-fold purpose of weakening enemy nations and increasing the population of Germany. It was also used as a method of retaliation and intimidation in occupied countries. During the war years numerous Czech, Polish, Yugoslavian and Norwegian children were taken from their parents or guardians and classified according to their 'racial value'.

I made a note of the countries identified in the charges: since the documents I had found in Gisela's papers indicated that I had been brought to the Reich for Germanisation, it seemed likely that I had originally come from Czechoslovakia, Poland, Yugoslavia or even Norway – though I somehow doubted that last was a real possibility. Perhaps there would be more solid clues in the rest of the papers in the Case 8 trial folder.

Before I could get to them, the list of defendants' names caught my eye. Four senior officials of Lebensborn – three men and one woman – had been in the dock at Nuremberg, their roles and ranks spelled out clearly.

MAX SOLLMAN – Standartenführer (colonel) in the SS; Chief of Lebensborn.

GREGOR EBNER – Oberführer (senior colonel) in the SS. Chief of the main health department of Lebensborn.

INGE VIERMETZ – Deputy chief of main department A of Lebensborn.

It was the fourth name that stopped me in my tracks.

> GUNTHER TESCH – Sturmbannführer (major) in the
> SS; Chief of the main Legal Department of Lebensborn.

I knew that name. Sturmbannführer Tesch had signed the document recording the contract under which I had been handed over to Hermann and Gisela. The man who had arranged my fostering had been indicted at Nuremberg for crimes against humanity.

I turned back to the disconcertingly large box of papers in front of me. The trial had lasted fifty-seven days, examined almost 2,000 exhibits and heard evidence from 116 witnesses for either prosecution or defence: the printed record ran to a staggering 4,780 pages. I began to wonder whether the three days I had planned to spend in Nuremberg would be enough.

The chief prosecutor, an American military lawyer called Telford Taylor, began by laying out the background to the charges. Just as I had learned in Hadamar, from the start Lebensborn's kidnapping and Germanisation programme had been an integral part of the Nazis' plan to eradicate those they regarded as 'lesser races'.

> With the launching of the wars of aggression by the
> Third Reich, it became possible to put these noxious
> principles into practice. By the middle of 1940, a very
> definite plan was being effectuated. This is shown by
> the top-secret document which Himmler wrote, entitled
> 'Reflections on The Treatment of Peoples of Alien Races
> in the East' …

I had never heard of this document – but then, as Brigadier General Taylor went on to tell the court, it was so secret that Himmler gave instructions that it was not to be copied or shown to anyone beyond a small inner circle of Hitler and the most senior party leaders. After quoting a section in which the Reichsführer pronounced his hope that 'the concept of Jews will be completely extinguished', Taylor read into the trial record a section of Himmler's blueprint for stealing those deemed pure Aryan children.

> The parents of such children of good blood will be given the choice to give away their child – they will then probably produce no more children so that the danger of this sub-human people of the East obtaining a class of leaders, which, since it would be equal to us, would also be dangerous for us, will disappear …
>
> If we acknowledge such a child as of our blood, the parents will be notified that the child will be sent to a school in Germany and that it will remain permanently in Germany.

Poland was the first country overrun by the Nazis. It became the testing ground for Lebensborn kidnappings. The court was shown a letter, dated 18 June 1941, in which Himmler spelled out his instructions.

> I would consider it right if small children of Polish families who show especially good racial characteristics were apprehended and educated by us in special children's institutions and children's homes, which must not be too large.

> After half a year the genealogical tree and documents
> of descent of those children who prove to be acceptable
> should be procured. After altogether one year it should
> be considered to give such children as foster children to
> childless families of good race.

Six months later he issued a new decree detailing how kidnappings were to work in practice, and how the wheat of
prospective 'racially valuable' children was to be separated from
the chaff of their parents – many of whom, he recognised,
would be active opponents of the Nazi occupation.

> Politically heavily incriminated persons will not be
> included in the resettlement action [the Reichsführer's
> euphemism for wholesale stealing of children]. Their
> names are also to be submitted by the Higher SS and
> police leaders to the competent State Police Main Office
> for the purpose of transfer to a concentration camp ...
> In such cases the children are to be separated from their
> parents ...
>
> The Higher SS and police leaders are to pay particular
> attention that the Germanisation of the children does not
> suffer as the result of detrimental influence by the parents.
>
> Should such detrimental influence be determined to
> exist, and should it be impossible to eliminate them through
> coercive measures by the State Police, accommodations
> are to be found for the children with families who are
> politically and ideologically above reproach and ready to
> take in the children as wards, without reservation and out
> of love for the good blood present in the children and to
> treat them as their own children.

Also included in the trial record were Himmler's directives on the fate of individual Polish families who did not meet the Nazi's standards for racial 'value':

> Brunhilde Muszynski is to be taken into protective custody. Her two children, aged four and seven years, are to be sterilised and lodged somewhere with foster parents.
>
> Ingeborg von Avenarius is also to be taken into protective custody. Her children too are to be lodged somewhere with foster parents, after sterilisation.

The court was presented with the transcript of a speech Himmler gave in October 1943, in which he justified these horrific orders:

> I consider that in dealing with members of a foreign country, especially some Slav nationality, we must not start from German points of view and we must not endow these people with decent German thoughts and logical conclusions of which they are not capable, but we must take them as they really are.
>
> Obviously in such a mixture of peoples there will always be some racially good types. Therefore I think that it is our duty to take their children with us, to remove them from their environment, if necessary by robbing or stealing them. Either we win over any good blood that we can use for ourselves, and give it a place in our people, or we destroy this blood.

Thus were the children of Poland – at least the blond, blue-eyed ones who conformed to the Nazis' belief in 'good Aryan characteristics' – snatched from their families and transported to holding camps. Here trained 'racial examiners' set up offices to measure, prod and evaluate thousands of youngsters brought to them by the SS. Their assessment was final and unchallengeable. A ruling issued by RuSHA explicitly stated:

> The racial sentence once passed ... by an expert may not
> be altered by any office. The judgment of an expert is an
> expert diagnosis just like that of a physician.

The chosen children were then handed over to Lebensborn. In the box of files there were records: closely typed sheaves of paper recording the names of those taken from their families and shipped to the network of Lebensborn homes across Germany. I closed my eyes and imagined the scene: railway stations crowded with thousands of unaccompanied children, stuffed like cattle into trucks and carriages, without anyone to care for them on their journey.

At that point, strangely, a fragment of memory came to me: one I had never remembered before and yet which somehow I felt to be true. I was very little; I was on a train, sitting on the floor with another small child. We were sharing a blanket, each of us trying to pull it from the other. I lost the battle and, as the train ploughed on through long, dark tunnels, I felt terribly cold. Was it real? Had reading the accounts of what had happened in Poland awakened a recollection that had lain dormant in my subconscious for sixty years?

The more I read in the Nuremberg files, the more uneasy and unsettled I felt. In my mind's eye I saw the trial unfolding, its focus shifting from Poland to a little village in what was then Czechoslovakia.

Marie Doležalová was fifteen when she stood on the witness
stand in 1947 and gave evidence about what had happened to
her five years earlier. On the morning of 9 June 1942, ten trucks
filled with SS and Gestapo troops rolled into Lidice, a farming
hamlet near Prague.

Two weeks earlier, Czech partisans had assassinated SS-
Obergruppenführer Reinhard Heydrich, Himmler's protégé
and the man charged with ruling this corner of the Reich's
conquered lands. Hitler demanded mass reprisals: the raid on
Lidice was specifically ordered because the village was suspected
of having links to the men who killed Heydrich.

Armed soldiers jumped out of their vehicles and rounded
up Lidice's entire population. Every adult man – 173 of them,
including Marie's father – was lined up and shot against the wall
of a barn. Their bodies were laid out in seventeen rows in an
orchard before the village was burned to the ground.

The women of Lidice, almost 200 in number, some heavily pregnant, were transported to Ravensbrück concentration
camp. Their children were snatched from them: 184 youngsters
were pushed into buses and transported to a former textile
factory in Łodz. On Himmler's staff orders they were not
fed, and were forced to sleep on cold dirt floors without
blankets.

Once RuSHA's 'race examiners' arrived in Łodz, they
assessed each child for signs of Aryan qualities. They 'failed' 103
children: of them, seventy-four were immediately handed over
to the Gestapo for onward transportation to the extermination
camp at Chelmno, seventy kilometres away. Here they were
gassed to death in specially adapted killing trucks. Just seven
children were selected as suitable candidates for Germanisation.
Marie Doležalová was one of them.

When she arrived at the children's home, she found herself among many other children from different countries. She was forced to learn German and was punished if she was caught speaking Czech. Lebensborn eventually handed her over to an approved German family. Her foster parents were kind, giving her two new dresses to mark her arrival with them, but she was encouraged to forget where she had come from.

After the war ended, the handful of Lidice women who had survived the massacre and the concentration camp began searching for their missing children. A year later – just before she gave evidence at Nuremberg – Marie was reunited with her mother, who was by then dying. As she stood at her mother's bedside, she realised that she couldn't remember a word of her native language.

All of this Marie Doležalová told the judges at Nuremberg. As I read her testimony, I put myself in her place. Perhaps I too had come from a village burned to the ground by Himmler's troops, one of the so-called 'racially valuable' lucky ones saved from the death camps by the promise of blond hair or blue eyes. But if so, where exactly had I been stolen from? And was there any hope that, like Marie, I might one day meet my real mother before she died?

And then I found the lists. They were tattered grey sheets of foolscap, prepared by Lebensborn staff in 1944; almost sixty years later the type had faded, making them only just legible. Each was divided into four columns. The first column was an alphabetical register of names and, since the adjacent column showed birth dates from the early 1940s, it was clear that this was some kind of register of children. The third column was headed 'transferred to', and in the final line there was a date against each entry.

There were 473 children identified in these documents. Halfway down one I read the following:

> Matko, Erika.
> [Born on] 11.11.41.
>   [Transferred to] Oberst Hermann von Oelhafen,
>       Munich, Gentzstrasse 5.
> [On] 3.6.44.

I had found my original name.

It had to be genuine: not only were these official court records but the address shown for Hermann – and the date on which I was handed over to him – was correct.

I sat back, the list in front of me. I was surprised to find that I was not emotional: ever since I had received the letter from the Slovenian government telling me that I could not be Erika Matko, because that person was still living in the Rogaška Slatina area, I had felt lost. Now as I looked at this fading Lebensborn list, I felt my purpose and true identity coming back to me.

Accompanying the lists were two sworn statements by former Lebensborn staff who had been interrogated by investigators for the Nuremberg prosecutors. The first was a woman called Maria-Martha Heinze-Wissede, who had worked in the Lebensborn head office. On 9 August 1948, she had been shown the documents and identified the origin of some of the children. Erika Matko was one of them.

> From the lists before me, I recognised the following names
> of Yugoslavian children...
>
> ... Erika MATKO

> These children, I know only their files a little, since they
> had already been transferred to ... German families by
> Lebensborn.
>
> As was clear from the documents, they were called
> 'bandit children', and Lebensborn took them over from
> Volksdeutsche Mittelstelle [VoMi] ... As far as I remember,
> the Lebensborn took these children from a Volksdeutsche
> Mittelstelle camp in the district of Bayreuth.

My heart raced. There it was in black and white: I had been
brought from Yugoslavia and passed to Lebensborn by this
VoMi organisation.

A quick search revealed VoMi to have been another of the
confusing and overlapping bodies answering to Himmler: it
was set up before the war, ostensibly to manage the interests
of the *Volksdeutsche* – ethnic Germans who lived outside the
borders of Nazi Germany. But once Hitler's armies overran
Poland, Czechoslovakia and Yugoslavia, VoMi settled half a
million German volunteers in the conquered territories, simultaneously shipping out or imprisoning the rightful occupants.
It was essentially a precursor to what we now call ethnic cleansing, and the organisation's involvement in my origins did not
bode well for the fate of my biological family.

Who were they? There was a clue, a thrilling hint,
in Maria-Martha Heinze-Wissede's affidavit. The other
Yugoslavian youngsters and I were listed as 'bandit children'.
In Nazi terminology this meant partisan fighters. I felt a surge
of pride: our fathers were rebels, they had opposed the Nazi
occupiers. How brave they must have been. I doubted that in
their position I would have found the courage to fight back
against Hitler's armies.

The second witness statement was from a former Lebensborn clerk called Emilie Edelmann. She had joined the organisation in 1939 and worked within it until the very end, rising to a position that gave her responsibility for the care of children being readied for placement with foster families. On 3 April 1948, she too told her American interrogators that children had been snatched from Yugoslavia, and filled in a few of the missing details of my journey to a Lebensborn home. She described these kidnapped children as *Südost-kinder* – VoMi-speak for those taken from the south-eastern territories conquered by Germany.

I re-read everything obsessively, desperate to be certain. It was unequivocal: here in the Nuremberg files was definitive proof that I had been one of at least twenty-five infants who in 1942 and 1943 were kidnapped and transported from Yugoslavia to the Fatherland. I had been taken to a VoMi holding camp at Werdenfels in southern Germany before being shipped on to the Sonnenwiese home at Kohren-Sahlis and then eventually given to the Oelhafen family in Munich.

I had only one question left to answer before I left Nuremberg: how did the trial end – what punishment was meted out to the senior Lebensborn officials in the dock?

Astonishingly, although most of the top RuSHA officials had been convicted and sentenced to lengthy terms in prison, the four Lebensborn defendants had been acquitted of crimes against humanity and war crimes. The three men had been found guilty of membership of the SS – as a woman, Inge Viermetz was excluded from its ranks – but none spent a single additional day behind bars. Despite all the evidence presented to them, the judges had reached the incredible conclusion that Lebensborn had been no more than 'a welfare organisation'.

I was furious. I had read the evidence and I had listened to the harrowing accounts from my fellow survivors at the meeting in Hadamar. I knew the truth now, and it made me more determined than ever to discover more about how I had originally fallen into Lebensborn's clutches.

Whatever the government of Slovenia thought, I had definitely come from there. I just had to prove it.

*'I would be very, very grateful if you could
answer some questions about your childhood ...
I am not asking out of mere curiosity ...'*
LETTER TO ERIKA MATKO, JUNE 2003

I found a large number of people throughout Germany with the surname Matko. I wrote to each address, asking if they knew anything about my background or that of Hermann and Gisela von Oelhafen. It was a succession of shots in the dark but, to my surprise, letters began trickling back: each thanked me for contacting them and wished me well but none of them were able to help with my investigation.

In the meantime, Josef Focks was busy. The 'Father Finder' was not deterred by the results of my letters to the German Matkos; instead he expanded his geographical search.

Josef Focks is one of the people without whom I would never have found the truth about my past. He was a former army officer who had been seconded to NATO in the 1980s. During a posting to Norway he first encountered the stories of children fathered by German troops and the plight of those who had been born in the Lebensborn programme. Moved by their pain and the sense of shame that had blighted their lives, he offered to help them track down their families.

From the outset, he ran into the problem that the Lebensborn officials had deliberately concealed many of the fathers' names. Ironically it was the logistical difficulty of genealogical searching in the days before the Internet and online records that led him to find innovative solutions. He made use of local contacts (he found that taxi drivers were good sources of information), dug into obscure archives and the libraries of old newspapers, and even visited cemeteries to examine the names carved into headstones. Gradually he developed a way to unlock the puzzle.

By the time I met him, he had taken on more than a thousand cases, not all of them Lebensborn children, successfully tracing family members in most of them. His investigations had led him across Germany and as far afield as America and Australia, and his office in Bonn was stuffed with innumerable files, each one bulging with paper. All of this he did without charging a penny for his time: he had long since retired from the army and lived on his state pension, helping people like me for nothing more than the reward of easing our pain. I cannot thank him enough.

It was Josef who found the most promising Matkos. He had tapped up one of his contacts, a woman whose mother had been taken from Yugoslavia by the Nazis as slave labour: with her help, he discovered contact details for several Matkos still living in or near Rogaška Slatina. They seemed to be an extended family: some were my age or slightly older, others clearly a generation younger. Most promising of all, one of them was named Erika.

Josef unearthed an address for her and also the telephone number for a Maria Matko, who, he thought, might be a relation. We agreed that he would phone Maria and that I would write to Erika.

I sat down at the computer and thought about what to say. It was not an easy letter to compose: I knew nothing about this woman nor the country in which she lived. In the end, I decided to speak openly about my need to discover the truth about my past. I told myself that since many of the unrelated Matkos in Germany who I had contacted out of the blue had taken the trouble to reply even though they could not help me, this person bearing my name and who lived in the place I knew I came from might also be moved by my plea for help.

Osnabrück, Germany

16 February, 2003

Dear Mrs Matko,

I am writing to you today about a very personal matter and hope that you can help me. The problem, of course, is that I do not speak Slovenian and I cannot hope that you speak German. But I hope that there is somebody who can help you to translate my letter.

For some years I have been researching my biological parents and during this research I found out very strange things, which have made me very anxious and disturbed, but I know that I must keep going on.

My foster parents picked me up from the Lebensborn home 'Sonnenwiese'. There I was given two vaccination documents in which my name is shown as Erika Matko, born in St Sauerbrunn. I don't have any other documentation about my early life. I don't know the circumstances which brought me to the home. My foster parents didn't give me any information.

Ten years ago I didn't even know that I was a Lebensborn child. The Red Cross couldn't find out any information about my identity. I asked Dr Georg Lilienthal, who is a researcher about Lebensborn and has published a book about the subject.

He gave me the idea that maybe I'm a member of the group of kidnapped children, and my origins lie in Yugoslavia.

Now in the course of my research I have discovered you. I don't ask from curiosity but I only want to know how this double identity has occurred. Did you ever live in a Lebensborn home or were you lucky enough to spend your whole life in Rogaška Slatina?

I would be very, very grateful if you could answer my questions about your childhood.

Best wishes, Ingrid von Oelhafen

There was nothing more I could say or do. I posted the letter, hoping that something in it would strike a chord with this other Erika Matko.

∽

In the meantime, Josef had made progress. He got in touch with Maria Matko and they had a good conversation via phone with the help of a translator, as he didn't speak Slovenian and she could not understand German. She was apparently my age and had spent her whole life in Rogaška Slatina.

From what Maria told Josef, she was the matriarch of the extended Matko family, which had once been involved with the anti-Nazi partisan movement. She was slightly hazy on the details, remembering only that one member of the family had been executed by the Nazis and that she had heard a story, long ago, that three children might have been kidnapped in the early 1940s. It sounded close to the likely Matko family history that I was looking for, and even more promising was the news that she knew the mysterious Erika very well.

But it was the final piece of information that threw me. Herr Focks had persuaded Maria that she should meet me – and that she should bring Erika with her. I was instantly nervous: I wanted desperately to go, but the prospect terrified me. What if these were the 'wrong' Matkos, and the trip turned out to be a wild goose chase? I would, I knew, be devastated. And even if these people were my relatives, that didn't mean a meeting would go well; perhaps they would be hostile or somehow resent me, which would be even worse.

The Father Finder was having none of it. He pushed and pushed until I agreed to his plan. This involved flying first to Munich, then on to Ljubljana, the Slovenian national capital. From there I would find a taxi to take me the eighty kilometres to Celje, the main town in the region.

There was an additional reason for going: every autumn a handful of survivors of the Nazis' kidnapping and deportation programme met in Celje. Josef had arranged for me to join them before heading on the next day to Rogaška Slatina, where I was to meet Maria in a cafe. Nor was I to go alone: he had asked a friend of his who spoke Slovenian to accompany me as my translator.

I knew precious little about Slovenia or its history at that point. I didn't even know how the country had come into being after the break-up of Yugoslavia. As the date for my departure approached, I started reading up, hoping to gain some insight into what life might have held for me had I not been stolen for the Lebensborn programme.

Yugoslavia had been one of the first countries to overthrow its German conquerors. Under the leadership of Josip Tito, the partisans were the most effective anti-Nazi resistance force in occupied Europe; by the middle of 1943 their activity had grown from running sporadic guerrilla raids to causing major military defeats and inflicting heavy casualties that Hitler's army could ill afford. By the start of 1944 they had managed to push the Wehrmacht out of the Serbian regions; a year later all German troops were expelled.

They achieved this with only limited support from the Soviet Union and although Tito's post-war regime was unashamedly communist – a single-party state with little tolerance of dissent or democracy – for the next twenty-five years the country was the most independent of all the Soviet Union's satellites behind the Iron Curtain. It began to distance itself from Moscow in 1948, determined instead to forge its own brand of socialism. It felt free to criticise the Kremlin and the West in equal measure and was one of the founders of the Non-Aligned Movement – the group of states which defiantly refused to ally themselves with either side in the Cold War.

But there were always tensions beneath the surface. The new nation was welded together from six separate and frequently hostile republics: Serbia, Croatia, Bosnia-Herzegovina, Macedonia, Montenegro and Slovenia. Each of these had very different ethnic, religious and political histories. What held them together was the inspirational figure of Josip Tito. His death in 1980 precipitated an unravelling of the whole country.

Serbs had always been the largest ethnic group in Yugoslavia and prior to the Second World War had been the most dominant force in the Kingdom of Yugoslavia. With Tito gone, Serbian communist leader Slobodan Milošević sought to restore this historic supremacy. The other republics, especially Slovenia and Croatia, denounced this power grab but were unable to stop it.

Industrial action by ethnic Albanian miners in Kosovo in 1989 was the spark that ignited the simmering tension. Slovenia and Croatia supported the Albanian miners and the strikes turned into widespread demonstrations demanding a Kosovan republic. This angered Serbia's leadership, which proceeded to use police force against the miners before sending in the Federal Army to restore order.

In January 1990, an extraordinary Congress of the League of Communists of Yugoslavia was convened. Since the country was a one-party state, this was effectively the ruling body for all of the Federal Republic. The meeting degenerated into an argument between Slovenia and Serbia about the future of the nation: in the end the League dissolved itself. The writing was on the wall for the future of Yugoslavia.

The immediate outcome was a constitutional crisis. Fuelled by a toxic rise in ethnic-based nationalism and inspired by the fall of communism across the rest of Eastern Europe, five of the republics demanded independence and an end to Serbian dominance. The stage was set for war.

What followed was Europe's worst conflict since the Second World War, and one that once again raised the spectre of crimes against humanity. Over the next decade, at least 140,000 people died in the fighting. Hundreds of thousands more – possibly millions – endured ethnic cleansing, rape as a weapon of war, concentration camps and mass bombing.

The first of these dirty wars broke out in Slovenia. In December 1990, 88 per cent of the population voted for full independence from the disintegrating federal republic, knowing that to do so would inevitably lead to an attempted invasion by the Serb-dominated Yugoslav People's Army. The fledgling Slovenian government secretly reorganised its antiquated territorial defence force into a well-trained and equipped guerrilla army and the partisan resistance which had once driven Hitler's troops from the country was effectively reborn.

The Slovenes knew that they stood no chance in a conventional battle: the YPA was simply too big and too powerful. So the country prepared for a campaign of guerilla warfare – a return to the resistance tactics of blowing up bridges and small close-quarter attacks in the towns and villages of their nascent nation.

At the same time, Slovenia sought help from the European Community and the United States. Neither was prepared to recognise the country's independence since they found it more convenient to deal with a single federation rather than a series of small states. The rebuff emboldened the Serbs and made a full-blown civil war inevitable.

The first shot was fired by the YPA on 27 June 1990 in the small village of Divača, just seventy-five kilometres from Ljubljana. That same afternoon, Slovenian soldiers shot down two Yugoslavian army helicopters.

Over the next ten days, the fighting moved westward towards Ljubljana, then on past the capital and into the eastern heartland around Celje and Rogaška Slatina. A ceasefire was announced on 6 July: Slovenia won its independence, though at the cost of at least sixty-two deaths and almost 330 wounded. By contrast with the ensuing conflicts in Croatia, Bosnia and

Kosovo, this was a small war, but it was the first time since the Nazis had been expelled that Slovenians had their freedom. In some rather inexplicable way, I felt proud.

∽

At the end of September 2003, I flew to Munich. Josef Focks had arranged for me to meet his translator at the airport so we could fly on to Ljubljana together. But by the time our flight was called he had not arrived and I boarded the plane alone. I was already nervous about who or what I would find in Slovenia and, since I spoke only German, I felt vulnerable and exposed. Fortunately the translator managed to get a message to the plane and asked a stewardess to tell me that he had been held up in traffic and would catch a later flight and meet me at Ljubljana.

I waited all day in the airport. I had no signal on my mobile phone, no one seemed to speak German and I could not work out how to use the local payphones. All I could do was sit and hope that my contact would turn up.

By the time he finally arrived it was mid-evening and I was in something of a state. But I had no time to dwell on my feelings: there was to be a meeting of the stolen Slovenian children that evening in the primary school at Celje. This town, I gathered, had been called Cilli during the German occupation, and had been both a centre for partisan resistance and the site of Nazi reprisals.

As we drove through the countryside, I looked out of the window, trying to take in the landscape, the land of my birth. I had wondered beforehand if seeing it for the first time in almost seventy years would prompt some memories: it felt disappointing to find that it did not.

I knew very little about the people I was meeting that evening and was surprised to discover that the event in Celje was a very different gathering of stolen children from my first encounter. They had started searching for one another as far back as 1962, determined to tell their stories to the (then) Yugoslavian public.

The men and women I met that evening were all in their eighties – between ten and fifteen years older than me. They were the leaders of what had become an officially supported group of survivors, and their accounts filled in some of the gaps in my knowledge.

Throughout 1942 a total of 654 children, from babies to eighteen-year-olds, were snatched from their families by the Nazis and shipped off to a succession of camps across the Reich. Most of the older ones – at least those who survived the rigours of slave labour or attempted Germanisation – had been brought back home at the end of the war. By the time I arrived in Celje, only around 200 were still alive.

Despite their age, their memories were strong and they were determined that the world should not forget what had been done to them. Two speakers stood up in the primary school to give testimony. I sat silently: even if I had felt able to trust the translator to speak for me, I knew too little to make any worthwhile contribution. But my presence had been noted. After the meeting was over, three people came over to speak to me. Each had been stolen from Celje in August 1942 and, to my complete astonishment, each one said they recognised me.

The first woman was seventeen when she was caught in a round-up of children from the area and held for two days in the primary school by SS troops. The children ranged from small infants to eighteen-year-old teenagers: because they were separated from their mothers, the older ones were ordered to look after the babies. This warm and emotional elderly woman told me that the babies were constantly crying. Her task had been to clean the smallest ones, and she specifically remembered washing me.

I couldn't believe what I was hearing. I would have been less than a year old when the round-up happened; how on earth could someone recognise me more than sixty years later? But somehow this woman believed that she did. It was astonishing. I had come to Slovenia hoping only to find some evidence that tied me to the country. Instead I had come face to face with someone who claimed that she could place me in Celje on the day that I was stolen – and who said she had actually held and looked after me when I was a baby.

On reflection it seemed more probable that the woman had been told I was coming and this had prompted her to remember me. But either way it was a connection.

The next person I spoke to was a man around the same age. He had been fourteen on the day of the kidnapping, and he was able to tell me a bit more about what had happened to the stolen children on the day after the round-up. He told me that we were transported 150 kilometres north to the holding camp at Frohnleiten in Austria. He was adamant that he had seen me there, and that the name he knew me by was Erika Matko.

Another elderly lady chimed in then, confirming what the man said. She had been kidnapped from Celje and shipped to Frohnleiten: she too remembered me there and that my name was Erika Matko.

I suddenly felt intensely happy. After so long, after so many disappointments, I had first-hand evidence of who I had once been and where I had come from. It was an extraordinary sensation.

I did not have time that night to press my new contacts for further details of the kidnapping. I would have to wait to find out more about the events of August 1942 and how I had been caught up in them.

∞

The next day, we moved on to the Maribor region. Ahead of my meeting with Maria, Josef Focks had arranged for me to meet two other people whose surname was Matko. He had also equipped me with sterile test tubes and cotton buds so that if these Matkos were willing, I could take samples of their saliva, which we would have analysed to see whether we shared any genetic similarities. In this way I could discover whether we were related.

Our first stop was a village near Rogaška Slatina. As we drove closer, I looked out of the window at the green and densely wooded landscape, hoping to see something familiar, but I didn't recognise anything. The village was clearly very poor: the woman I was due to meet was eighty and lived with her forty-year-old son. Both seemed puzzled by my visit but the old lady agreed to give me a saliva sample. Her son was more hostile and refused. Neither of them were able to tell me much about their family background and I left thinking that if we were related it was probably only distantly.

The next Matko was a thirty-year-old hairdresser. She willingly gave me a sample for DNA analysis but, again, had little information to help me on my journey.

Finally it was time to meet Maria Matko. Herr Focks had arranged our rendezvous in a little cafe in Rogaška Slatina, and she had promised to bring the mysterious other Erika with her. But the moment I walked in, I could see that Maria was alone. I felt a sharp pang of disappointment.

Maria herself turned out to be warm and helpful, and a vital link in unlocking the chains of my past. She was seventy-three years old and not a Matko by birth; she had married into the family. Her husband was called Ludvig and he had two sisters: Tanja, who was older than him, and Erika, who was younger. Both Ludvig and Tanja were dead, but Erika was still alive, if rather infirm. Maria said that in the end she had been unwilling to meet me.

It was plain from our conversation that none of the extended Matko family thought I was related to them. Maria was the most open-minded, but even she was sceptical. At that point I was beginning to share their doubts.

But then Maria gave me the details that once again raised my hopes. The parents of Ludvig, Tanja and Erika had been called Johann and Helena – the very names I had been given three years before by the archivist in Maribor. What's more, Johann had been imprisoned by the Nazis for resistance activity: that seemed to fit what Georg Lilienthal had told me about my background.

Over the preceding years there had been a pattern to all my attempts to discover the truth: one piece of new information would emerge to lift my spirits and make me believe that I could find the answers I sought. But it would always be followed by a letter or conversation that dashed those hopes and sent me back into despair. And so it proved that day in Rogaška Slatina: no sooner had my expectations been raised than they were lowered once again. Maria produced two photographs. One was of Helena, taken in 1964: she was looking directly at the camera, a solid-looking woman with dark hair and a strong but kindly face. The second picture was of Erika, and my immediate reaction was that she looked just like Helena. That suggested she was Helena's daughter – and if so, then surely it meant that I could not have been.

When we parted, I felt a little depressed. The only thing buoying my spirits was that Maria agreed to meet me again, and promised to talk to the other members of the Matko family to see if they would cooperate with me.

∽

The next day I went to her apartment. She was babysitting her granddaughter, and talked openly about the family history. She told me that Ludvig, Tanja and Erika had lived with their parents until Johann was arrested by the Nazis and imprisoned in a concentration camp. In the early summer of 1942 he had been released and returned to the family home, after which point her knowledge ended. Johann's brother, Ignaz, had not been so lucky: he too was a partisan and arrested by the Germans, but he was shot by a firing squad.

During our conversation, Maria's son Rafael and her nephews Marko and Pieter joined us. Rafael was in his forties, solidly built and balding. Although he and his cousins were polite, they clearly did not believe I was related to them. It was obvious that the family was close-knit and protective of one another.

The one exception was, apparently, Erika. Maria told me that her sister-in-law had never married, though she did have a son, and had been so ill throughout her life that she had never worked. It was plain that for all the family bonds – Erika was usually invited to Sunday lunches – the two women weren't particularly close.

By the time I took my leave, I had convinced myself that although this family seemed to share some remarkable similarities with the Matkos I was searching for, they were probably not my blood family. Nonetheless, before I set off I explained about my DNA testing plans. Rafael kindly agreed to give me a saliva sample and, after some hesitation, so did his cousin Marko.

My final appointment was in Maribor. I went to the museum, which maintained a special section dedicated to the remembrance of what the Nazis had done in Slovenia. In addition to rounding up and executing 'bandits' – partisan fighters like Ignaz and Johann Matko – all Slovenian books were burned, the language was banned and anyone caught speaking it was severely punished. Himmler's plan to subjugate the local population was put into effect brutally and efficiently.

Perhaps once I would have been shocked by what I saw in the museum, but everything I had learned over the previous three years about Lebensborn and its operations had inured me to the routine cruelty of the Nazi occupation. Instead I was preoccupied by trying to process what I had discovered during my four days in Slovenia. I had arrived uncertain about how I fitted into the country's history in general, and into the story of the Matko family in particular. With each contradictory new piece of information I had, by turns, been convinced that I was Erika Matko from Rogaška Slatina and then certain that I could not be her. By the time I got back on the plane in Ljubljana, I was completely confused.

I had been away from my physiotherapy practice in Osnabrück for too long. My patients needed me, and I needed them. It was time to go home: to be Ingrid von Oelhafen once again.

*'Scientific analysis is 93.3 per cent certain ...'*
DNA TEST RESULTS, OCTOBER 2003

How do we put together the jigsaw puzzle of our lives? It is not easy, even when none of the pieces are missing and we have the picture on the box to compare it with. How much harder when we lack the solid, clear-cut corners to build out from.

This was how the story of my childhood looked to me when I arrived home from Slovenia. I had dozens of individual pieces but they were oddly shaped, sometimes overlapping, sometimes contradictory. There seemed to be no way to fit them together to reveal the full picture.

I couldn't decide whether or not to believe the survivors of the Celje kidnappings, so convinced that I was the Erika Matko they remembered. Surely it wasn't possible that someone would recognise in the face of a sixty-two-year-old woman the features of a nine-month-old baby they had known only in the most traumatic of circumstances? Plus their accounts were at odds with the reaction of Maria Matko and her son: they plainly did not believe I was part of their family. And always in the background was the shadowy figure of the other Erika, who had still not replied to my letter and who had deliberately avoided meeting me in Rogaška Slatina.

The answer, I hoped, lay in the swabs and test tubes I had brought home with me. Saliva analysis would provide genetic fingerprints, which are as reliable as a blood test in determining family relationships. The irony of this was not lost on me: the Lebensborn experiment had been based on the Nazis' belief in blood as the determining factor of human worth. Himmler's obsession with blood and bloodlines was the reason I had been plucked from my family – whoever they were – in Yugoslavia and reborn as a German child. It had shaped the course of my life from that day onward. Now I was trying to use it to unravel the tangled web Lebensborn had spun.

The saliva swabs had not been given without a great deal of anxiety. There had been a dispute within the Matko family about me taking them: some of the younger members of the family had been adamantly opposed to the idea of scientific tests to establish whether I was one of them. They worried, I think, primarily about the stress this could cause to Erika, who was not in good health. Others, though, believed that it was important to find out one way or the other. After much discussion, I had been given the family's blessing to have their samples analysed. I set about finding a laboratory to test them for me.

Only at this point did I discover that it was not going to be easy – or cheap. The science of DNA testing really began in 1985 and was then both rudimentary and the exclusive domain of law enforcement authorities. Although it had since been refined and made more commercially available, it was still expensive.

All of the cotton buds I had collected in Slovenia had to be processed so that the individual DNA of each person could be isolated; I also gave a sample for comparison. Although

99.9 per cent of human DNA sequences are the same for everyone, there are enough differences to make it possible to distinguish one individual from another. What the scientists would look for were places in these sequences called 'loci'. Where two people are related by blood, these loci are very similar to each other; in samples from biologically unrelated donors, they look completely different.

I used my savings to pay for the tests. It meant tightening my belt and not taking holidays for the foreseeable future, but there was simply no alternative. I carefully parcelled up the swabs and sent them off to the laboratory in Munich. It took several months for the results to come back. What they revealed was both the answer to the big puzzle and a new mystery.

I looked first at the analysis of the samples provided by the hairdresser and the old lady. As I had suspected, neither had any genetic relationship to me. I allowed myself a guilty feeling of relief: it was clear that the old lady lived in some poverty, which was sad to see. All along I had hoped that my biological family had not suffered; it was painful to imagine that my real mother might have had such a hard life.

The next set of results were for Rafael Matko, the son of Ludvig and Maria. When I read them, my guilty feelings were transformed into pure happiness.

> Scientific analysis shows that Ingrid von Oelhafen and Rafael Matko are relatives in the second degree ... it is 93.3 per cent certain that Ingrid von Oelhafen is the aunt of Rafael Matko.

There it was – the evidence I had been seeking for so long. If I was Rafael's aunt, that meant I was Ludvig's sister, and therefore the daughter of Johann and Helena Matko. I had found the vital piece of the jigsaw puzzle: I was unquestionably Erika Matko from St Sauerbrunn/Rogaška Slatina.

It is hard to convey what this news meant to me. Unless you have lived your life as I had, haunted by never knowing who I was and where I came from, I don't think it is possible to fully understand the overwhelming elation: it was like being liberated. I felt as though the weight of sixty years had been lifted from me.

Then, as always seemed to happen, the other test results dragged me back into uncertainty. Analysis of the saliva sample given by Marko Matko – Rafael's cousin – had yielded what appeared to be a completely contradictory outcome: it showed that to a 98.8 per cent degree of certainty I was not related to Marko.

This simply didn't make sense. I looked again at the Matko family tree to remind myself of what I already knew: Johann and Helena had three children, Tanja, Ludvig and Erika. Ludvig's son was Rafael and the tests proved that I was his aunt: therefore I must be Erika. But the same set of results showed that Tanja's son, Marko, was not my biological relative. No matter how I arranged these jigsaw pieces, I could not get them to fit. If I was Ludvig and Tanja's sister, why wasn't I Marko's aunt? The Matko family seemed to be surrounded by secrets.

The person most likely to have the answers was the mysterious other Erika. She was the last living first-degree blood relative – at least in theory. She had been raised by Helena and Johann and had grown up with Tanja and Ludvig. But she was still ignoring me and I had to assume by this point that she was not prepared to help me. It was immensely frustrating; I could not understand why she was being so obstructive.

I tried to focus on the positives. I knew for certain that I was – or had once been – Erika Matko, daughter of Johann and Helena, and that Ludvig at least was my brother. Quite how Tanja and the other Erika fitted into the picture was still a mystery, but at least I could be certain who my biological parents were. That was a very real comfort. But I was still no closer to solving the puzzle of how I had been taken from them. It would be another four years before those pieces fell into place.

'To be rooted is perhaps the most important and
least recognised need of the human soul.'
SIMONE WEIL, THE NEED FOR ROOTS, 1949

Wernigerode lies on the eastern edge of the Harz
Mountains in the German heartland of Saxony. It
is a quiet, picturesque town, in which half-timbered houses
flank the Holtemme river and horse-drawn carts still rumble
through cobbled streets. It looks and feels like the setting for a
fairy tale: the sort of place where the Brothers Grimm might
have based their stories.

But Wernigerode has another, altogether less cosy history.
On the top of a steep hill just outside the town lie the ruins of
Heim Harz, one of the network of Lebensborn homes.

In late summer 2005, I drove to Wernigerode to take part
in the creation of a new organisation. Lebensspuren ('Traces
of Life') was the first formal attempt by those of us who had
been born or reared under Himmler's Master Race programme
to band together: our aim was both to provide much-needed
support and to begin the process of bringing Lebensborn
into the public gaze, free from the prejudice and shame that

prevented others from understanding what had been done to us in its homes.

It was a long journey. The road stretched more than 260 kilometres through the woods and fields of central Germany: as I drove I had time to reflect on how I had got here. It was more than five years since I had begun the search for my roots. I had learned so much during that time, and yet I still knew relatively little.

In the ten months since I had received the scientific tests that proved who I was, I had made no real progress in discovering how I had been brought into the Lebensborn programme, nor did I fully understand the extent of the experiment itself. In this I was far from alone. The meeting in Hadamar had been a first step for the handful of Lebensborn children to come together and share our stories. Each of us had a little part of the overall jigsaw, but even together we could not complete the full picture.

The title of our new organisation was a deliberate twist on Lebensborn: where that had been, in Himmler's vision and language, the Fount of Life, our association was to be the way for its survivors to explain it. But I was also aware of a possible play on words: that middle syllable, 'pur', was an acknowledgement of the Nazis' obsession with racial purity which lay at the root of all our problems.

We chose a particular quotation to head our articles of association:

Uprootedness is by far the most dangerous disease to which human society is exposed. Whoever is uprooted, uproots others. Who is rooted himself, doesn't uproot others. To be rooted is perhaps the most important and least recognised need of the human soul.

It came from the French philosopher and activist, Simone Weil. She had fought fascism in Germany in the early 1930s and later as a volunteer on the Republican side during the Spanish Civil War. In 1943 she wrote a book called *The Need for Roots*, examining the social, cultural and spiritual malaise undermining western society; the quotation we selected from it perfectly encapsulated the story of our lives.

∽∞∾

I liked Guntram Weber the moment I met him at the Lebensspuren meeting. We were staying in the same guesthouse and we shared a common interest in working with young people: Guntram was a creative writing teacher specialising in helping disadvantaged youngsters. He was two years younger than me, but his face bore witness to the pain he had experienced throughout his life. When he stood up to tell his story, his eyes filled with tears as he described his struggle to find the truth about his origins as well as the overwhelming desire to run away from it.

He had grown up in an outwardly normal post-war German family, living with his parents and two siblings – an older sister and a younger brother. But behind closed doors it was a different story.

> As a child I remember sensing that I wasn't quite normal. Relatives seemed to treat me awkwardly and it gradually became clear that the man I called Father was actually my stepfather. Of course I wanted to find out who my real father was, but the subject was taboo in our house.

Relatives had been well drilled by my mother to hide
the truth behind vague statements. 'It was the war,' they
would say. 'Things were very confusing. We didn't see
much of each other – you will have to ask your mother.'

It wasn't until he was thirteen that Guntram's mother agreed
to discuss the issue.

'Well, Guntram,' she said, 'You are old enough to know
the truth about your father now.' Then she gave me a
name, told me when his birthday was and that she had
married him in 1938 on a beautiful sunny day and that
they had driven to church in a horse-drawn cart.

During the war he had been a truck driver for the
Luftwaffe, far away from the front, who had died driving
over a landmine in Yugoslavia. She added that he certainly
wasn't involved in killing anyone.

But there were no documents and no photos of this
man, and when I pressed her, my mother she said she
didn't want to say any more about him because it was
too painful.

It was a plausible story. Guntram was a little suspicious, but the
climate of German society during the 1950s actively discouraged awkward questions. Many children were told lies about
what their parents did in the war and, as I knew from my own
experience, it wasn't the 'done thing' to challenge them.

Curiosity and uncertainty gnawed away at Guntram.
Sometimes he thought about confronting his mother about
his doubts, but he never managed to do so. The lack of any
photos or documents led him to question the story of the

non-combatant Luftwaffe driver: instead he began worrying that his father had been a Nazi and that this was the reason for his family's secretive behaviour. He began inspecting his facial features in the mirror and poring over history books in the school library, searching for photos of soldiers who could be his father or for women concentration camp guards who looked like his mother. For one terrible period he even convinced himself that Joseph Goebbels, Reich Minister for Propaganda and one of Hitler's most devoted followers, might be his father. A year or so later he made a disturbing discovery.

> My mother had a strongbox in the bottom right-hand corner of her wardrobe. One afternoon when she was out, I decided to look in. I had terrible qualms about doing this; I knew I was breaking the trust between us and she was my only security in the world. But I felt I had no choice.

Inside the trunk, Guntram found the first clue to his identity: a small silver cup. It bore a profoundly unsettling inscription.

> We were a fairly poor family at the time. Like many others, my mother had lost everything during the war, so to find a silver object in the house was extremely unusual. I picked it up carefully and discovered my Christian name on it; my second name, though, was shown as 'Heinrich'. And then I turned it over and saw the writing on the other side. It read: 'From your godfather, Heinrich Himmler.'

Guntram desperately wanted to ask his mother about the cup. But, like Gisela with me, she was secretive – and he knew how badly she would react to the revelation that he had been rummaging through her things. It remained an unspoken and unsettling mystery.

In 1966 he first heard the word Lebensborn. His older sister needed her birth certificate in order to get married and was surprised to discover that she didn't have one. When she questioned their mother she was obstructive and told her daughter she didn't know where it was.

An enquiry at her place of birth turned up the unexpected news that Guntram's sister was the illegitimate child of an army officer. Her records were still intact and showed that she had been born in a Lebensborn home. That led to the revelation that Guntram had also been a Lebensborn child. Rather than question his mother further though, he chose instead to get away, and moved to the United States. He stayed there for eight years and had a family of his own, putting aside questions about his past.

But when his partner died in a car crash, he returned to Germany with his son. Before long the uncertainty about his roots began nagging at him again and finally, in 1982, he decided to confront his mother during a long car journey where, as he put it, 'she could not escape from me'. He pulled off the road and forced his mother to talk to him.

My mother was angry but she uttered three sentences that I will never forget. First she said: 'I don't want to talk about that.' Then she tried to stop me digging into my past: 'People will throw dirt at you,' she told me. Eventually, when she saw that I would not be put off, she made a promise to write the whole story down for me. I believed her and felt better, trusting that she would give me the truth.

Sadly she never did. For Guntram's mother the truth was simply
too difficult to speak about: she died two years later, taking her
secrets to the grave and leaving Guntram both frustrated and
angry. As she had once told him in an unguarded moment:
'The relationship between mother and child is a power struggle.'
Guntram felt he had been powerless in that struggle.

It wasn't until 2001, when he was fifty-eight, that Guntram
discovered who his father was – not, as his mother had told him,
a young soldier who died honourably, but an SS major-general
who oversaw the deaths of tens of thousands of people while
stationed in what is now western Poland. He had been convicted of war crimes and sentenced to death by a Polish court
in 1949, but had escaped to Argentina, where he died in 1970.

> My father was a war criminal. He was a man who allowed
> himself everything. And the SS enabled him to live that
> way. I assume my mother fell in love with a powerful
> military man.
>
> He died peacefully and at his funeral his old com-
> rades stood beside his grave and raised their right arms
> in the Nazi salute. I knew then that a racist is always a
> racist.

There was a bitter irony to Guntram's description of his life.
As a Lebensborn baby, his 'racially pure' genes were supposed to have ensured that he grew up strong and confident
– a future leader of the Master Race. Instead he had suffered
for more than sixty years from feelings of low self-esteem,
loneliness and uncertainty. The only thing that helped, he
told us in Wernigerode, was finding other Lebensborn
children.

It has been a huge relief for me, although I haven't been able to shake this feeling of inadequacy. Maybe in ten years it will be gone. It's important that other children in Germany and abroad hear about this group because it could help us all.

I agreed wholeheartedly with Guntram: Lebensspuren needed to be publicly visible so that other men and women who had been through the Lebensborn programme could make contact with us and perhaps find some solace. But in 2005 our group was not ready to do that: we met privately – partly due to the sense of shame still attached to our past.

To some extent Helga Kahrau exemplified the dichotomy we all faced, that of needing support and acceptance while simultaneously struggling with the painful reality of her birth. A tall and forceful woman, with her blond hair dyed a striking red, Helga had been born into the heart of the Nazi regime. During the war her mother, Margarete, had been a secretary in the offices of Hitler's top aide, Martin Bormann, and of Joseph Goebbels. She had memories of growing up in privilege and comfort, surrounded by important-looking men in crisp uniforms.

In the decades that followed the end of the Third Reich, Margarete refused to talk to Helga about the war, much less tell her about the father she had never known. It was only after Margarete's death in 1993 that Helga began to examine her family's past. She was horrified by what she discovered.

Margarete was a fervent Nazi who barely knew Helga's father. He was a German army officer and they met in June 1940 at a party celebrating Hitler's conquest of France. They had a one-night stand, which left Margarete pregnant. She was a perfect candidate for Himmler's Master Race programme:

politically committed, racially pure and expecting an illegitimate child from an equally Aryan German soldier. Nine months later, Helga was born in the main Lebensborn home at Steinhöring, outside Munich.

When Helga was three months old, Margarete left the home and returned to work in Goebbels' Propaganda Ministry. Helga was handed over to foster parents: her new father was a high-ranking Nazi official in the occupied Polish city of Łodz. Here, Helga believed, he helped oversee the gassing of thousands of Jews at the nearby Chelmno concentration camp.

I spent the first four years of my life raised and tutored by the Nazi elite. I was involved, in a fundamental way, with murderers.

At the end of the war, unlike many Lebensborn children, Helga was sent back to Munich to live with Margarete. Here she encountered the irony of Himmler's obsession with Nordic features. Although the city and its surrounding regions were the birthplace of Nazism, most Bavarians are dark-haired: the very racial characteristics which Lebensborn valued ensured that Helga stuck out.

I was big, blond and Aryan, different from the southern Germans, and everyone asked me where I had come from. I couldn't answer them.

The only document Helga possessed was a cryptic birth certificate from an 'SS Mothers Home'. It showed her mother's name but not that of her father. Nor was Margarete Kahrau willing to help her daughter understand. She deliberately concealed the truth, saying only that Helga's father had been a soldier who died in the war. She was also reluctant to discuss what she had done in the service of Hitler's Reich: like most Germans of her generation, Margarete preferred to forget about the Nazis.

When Margarete died in 1993, Helga began investigating. She discovered Nazi files that provided detailed information about her foster father and the crimes he committed in the service of the Final Solution. But there was nothing in the documents about her biological father. Then, in 1994, she received a phone call. The man told her that he had been a Wehrmacht officer in Paris, that he had met Margarete for a single night of passion. He was now suffering from terminal cancer and wanted to reach out to his daughter. It was a bittersweet moment for Helga: she had found her real father at last, but only as he was dying. She decided to make the most of the time they had left and devoted herself to nursing him round the clock.

Her father had enjoyed a very successful post-war career in property, which had made him a multi-millionaire. As his eldest child, Helga might have expected to inherit at least some of his estate. But when he died she encountered another legacy of her birth. Her father left no will: shortly after his funeral Helga received a letter from his lawyers stating that because she was illegitimate, she could legally inherit nothing.

Since then, Helga had found some solace in visiting her birthplace, the Lebensborn home at Steinhöring. But she never came to terms with her identity and worried constantly that people would assume that she, like her mother and stepfather, was a Nazi.

I grew up on the side of the murderers. Being a Lebensborn child is still a source of shame.

Shame – the word that has blighted the lives of so many of those
who had been part of Himmler's plan to create a new Master
Race. The more I heard from those who had been born into
Lebensborn, rather than kidnapped to strengthen it, the more
I felt lucky to have been one of the *Banditenkinder*, the child of
courageous partisans who fought against Nazi rule.

Gisela Heidenreich, a tall, strikingly Aryan-looking woman
from Bavaria, was four years old when she first encountered
shame: she overheard her uncle describe her as 'an SS bas-
tard'. She was a family therapist, a fact I noted with interest.
There seemed to be a common trait in Lebensborn children:
by accident or design many of us had chosen careers in which
we helped others overcome problems, while struggling with
our own.

Gisela described vividly the confusion she had suffered
throughout her life and the web of lies that blighted her childhood. Her mother was Emilie Edelmann, the Lebensborn
secretary who had given evidence at Nuremberg and who had
been responsible for finding foster parents to take children stolen from the occupied countries.

Like so many of those who had been part of Lebensborn,
Emilie was secretive and deceitful. At first she had led Gisela
to believe that she was not her real mother but her aunt. Later
she admitted that this was not true and told her daughter that
she fell pregnant following an affair with a married man. The
SS packed her off to occupied Norway, to give birth in the
Lebensborn home near Oslo. Several months later, Emilie
brought Gisela back to Germany.

Throughout her childhood, Gisela's mother refused to
answer any questions about the war. Only after Emilie's death
did Gisela discover the depths of her mother's involvement with

the Nazis. She found a bundle of love letters written to Emilie by Horst Wagner, the director of 'Jewish Affairs' in the Reich's Foreign Office and its link-man with the SS. In this role, he helped carry out the round-ups, deportation and extermination of both German and foreign Jews. Emilie's love affair with Wagner grew to the point where the couple considered formally making him Gisela's stepfather. Their relationship continued when the Reich fell and he was arrested by the Americans and held for trial at Nuremberg. But before he could be brought to justice, he fled down one of the infamous 'Ratlines' – clandestine escape routes for Nazi war criminals – to South America.

Shocking though the relationship was, it did not bring Gisela any closer to discovering who her real father was. She continued to search for him and, many years later, she eventually tracked him down. He was the head of the SS officer school at Bad Toelz in Bavaria. Her own reaction to their reunion both surprised her and helped her understand how so many Germans were able to live with knowledge of the crimes committed by the Nazis:

> When I first met him it was on a station platform. I ran into his arms and all I thought was 'I've got a father'. In that instant I sanitised the person I knew my father was. And I never asked him what he did. My own reaction – that of an educated adult with knowledge of the Lebensborn programme – has helped me to understand how people in those days just put the blinders on and ignored the terrible things that were happening.

Gisela brought to our group a determination to rehabilitate the image of Lebensborn children. In part this was due to her experience of the post-war treatment of Norwegian babies born in the programme's homes. As I had heard at the first meeting in Hadamar, Norwegian hatred of the occupying German armies led to discrimination against the 8,000 children like Gisela born in its Lebensborn homes. At first the post-war government in Oslo tried to have all the children shipped to Germany. When that plan failed, many of them were locked away in mental institutions or children's homes.

Gisela believed that this hatred and persecution was driven by Norway's national guilt at being occupied, the shame of its leaders having collaborated with the Nazis and, above all, the wildly inaccurate rumours about Lebensborn homes being 'SS stud farms'. Three years earlier, the Norwegian government had quietly paid on average €24,000 compensation to each of the children it had victimised. Now, Gisela argued, the time had come to end the lies and the discrimination.

> It's high time to tell the truth. There's been too much talk about Nazi babies, women being kept as SS whores and tall, blond people being bred. The Holocaust was about extinguishing so-called lesser races. Lebensborn was the reverse side of this coin: the idea was to further the Aryan race by whatever means were available.
>
> What I have learned is that I, and every other Lebensborn child, have a feeling of deep uncertainty about our identity. This has to stop.

The stories of these 'pure' Aryan children were harrowing. But theirs was only half the picture: there were others, like me, at the inaugural Lebensspuren meeting who had been forced, not born, into Himmler's programme. Their accounts helped me understand how the process had worked.

What happened to Barbara Paciorkiewicz was typical. She was born in 1938 in Gdynia, near Gdansk in Poland. Her family name was Gajzler but, because her mother had died and her father had disappeared, she and her sister were separated, each sent to live with one of their grandparents.

Gdansk was in the part of Poland under German occupation and the Nazis had renamed it Danzig. In 1942, when Barbara was four, the Youth Welfare Office issued instructions for all children to be brought to the regional youth welfare office in Łodz. Her grandmother took Barbara and was forced to leave her there.

There were a lot of children in the centre. Each was measured – heads, chests and hips – and weighed on scales. Their faces were photographed from three angles. The people making these measurements were some of Himmler's race examiners: they were looking for racially suitable children to Germanise. Barbara had blond hair and looked Nordic. She was shipped off to a succession of different homes in Łodz.

> Here I was subjected to more tests – always there were
> more tests. We were forbidden upon pain of punishment
> to speak Polish. Every one of us was crying.

Barbara's account of her kidnapping was similar to others I had heard. But her recollections of life in the Lebensborn home at Bad Polzin gave me new information about what my own experience might have been like.

> This is where my memories truly begin. I can remember
> exactly where we were kept, the conditions there, and the
> treatment we received. There was a separation between

us stolen children and those who had been born in the home.

The stolen children were kept on the ground floor of the building. The babies born in the home were kept on the floor above and we were never allowed to mix with them; nor were the staff who looked after these babies allowed to mix with the staff in charge of us. It was as if there was a hierarchy of our value: apparently the Nazis viewed the babies as more important than us children brought in for Germanisation.

In the home we were constantly given medical tests – I think it must have been every day. We were in a big room on the ground floor which had a large semi-circular wall of windows. I have been there since and this room still exists: it looks almost the same. There was a very sinister atmosphere in the room and we were individually taken into a side room and given injections by a doctor. I fear now that these were to tranquillise us: I cannot see any other reason for it. We were terrified of these injections. All the children were crying in that room: no one ever laughed.

Even for the precious Aryan babies like Guntram Weber, the regime in Lebensborn homes was cold to the point of severity. They were separated from their mothers immediately after birth, and kept apart for the next twenty-four hours. Thereafter they were allowed just twenty minutes together every four hours: even during that brief period of contact the SS staff strongly discouraged mothers from caressing or talking to their children.

The older children were monitored constantly, and reports made about their behaviour. Uncleanliness, bedwetting, farting, nail-biting and masturbation (which older boys were told on arrival was forbidden) were enough to ensure expulsion: these rejects were shipped off to forced education camps where they were brutalised or used as slave labour.

This Spartan regimen was intended to produce strong and ruthless future leaders for the Master Race. But children need love, not unyielding discipline: Barbara Paciorkiewicz remembered clearly how the rules frequently produced the opposite effect to Himmler's objective.

> The children often reacted to them by wetting their beds.
> In the mornings when this was discovered, the children
> were beaten for this: even if only one had wet the bed, all
> of us were punished.

Himmler's overall plan was the same for both types of Lebensborn children: wherever possible, for the duration of the war, they were to be handed over to carefully vetted foster parents who would raise them as model Aryans. After Germany's eventual victory, the boys were to be sent to elite schools – the network of SS-run *Nationalpolitische Erziehungsanstalten* – where they would receive a strong physical and political education. The girls were to be sent to schools run by the Bund Deutscher Mädel (the female equivalent of the Hitler Youth) and trained to become housewives and mothers.

Barbara had also uncovered the way in which Lebensborn deliberately obscured the true identities of children stolen from the occupied territories. First it cut an essential link to their home country by forbidding them to speak their own language; then it told would-be foster parents that the children were the orphans of fallen German soldiers. The men who

ran the programme – those who had been indicted and then
acquitted at the Nuremberg trial – knew this to be a lie but
Himmler's orders made it clear that for youngsters like Barbara
and me, every single trace of our previous life in Poland or
Yugoslavia was to be erased.

Barbara's foster parents were from Lemgo in North
Rhine-Westphalia. The Rossmanns were in their fifties and had
two grown-up sons who had been drafted into the Wehrmacht:
they had also had a daughter who had died of scarlet fever when
she was nine. Neither was, as far as Barbara had been able to
discover, a member of the Nazi Party. Herr Rossman was the
director of a school and his wife was a housewife.

Although they were good and kind people who longed for
a child to replace the daughter they had lost, even as a young
child Barbara felt out of place in her new family. She suffered
from bad nightmares in which unknown men came in through
an open window to steal her.

> There was always an air of uneasiness at home. When I
> entered a room, everyone stopped talking. And I always
> asked myself: 'What is it about me that makes this
> happen?'

Back in Poland, Barbara's grandmother had never given up hope
of finding her. At the end of the war she contacted the Red
Cross: it located documents showing where Barbara was living.
Not long after, she was taken from her foster family and placed
in a temporary children's home run by the British Army. Six
months later, she was put on a train to Poland. She was eleven
years old and had never been told about her biological parents,
nor that she was anything other than a normal German child.

I was very frightened and confused: I still thought the Rossmanns were my real parents and I didn't know anything about Poland. It didn't mean anything to me. I didn't speak Polish and I didn't even know I had a grandmother. It all seemed like a terrible journey into the unknown.

I realised that I had never given any thought to what had happened to other stolen children after the end of the war. It had never occurred to me that some children might have been traced and sent back to a country they could not remember. Barbara's story made me wonder whether, had I been given the choice, I would have wanted to be returned to Yugoslavia.

The journey to Poland took a terribly long time. The train sometimes stood for days in a siding. And then, suddenly, I heard people around me shout out 'Poland, Poland' and everyone was happy. I felt no joy when I arrived.

We were taken to a Red Cross camp at Katowice. It was chaos with people running around shouting out names. My grandmother had sent my uncle and he was shouting, but I had no idea who he was or what he was saying because he was speaking Polish. It was only when he shouted 'Gajzler/Rossmann' that I realised. And I also realised at that moment that I had completely lost my identity.

In the ruins of post-war Poland, a country with every reason to hate Germany, Barbara felt confused and isolated. Her uncle and his wife had been in a Nazi forced labour camp together with older children stolen from Gdynia but who were not deemed racially valuable.

Barbara was taken to visit the former Stutthof concentration camp near Gdansk. Her uncle forced her to look at the piles of children's shoes and the gas chambers: 85,000 people had died here and he wanted her to see and understand what the Germans had done. But she still thought of herself as German. She found it impossible to believe what her uncle said.

> I thought German people were all good: that was how I had been brought up. It was even worse in school: the children played games in the playground in which Hitler was the bad guy. Because I was German, I was always picked as Hitler: but I didn't mind – in fact, in my ignorance, I shouted out with pride that Hitler was my uncle.
>
> I always thought that a terrible mistake had been made: that I had been mistaken for someone else, and this was why a good German girl like me was in this strange place.

Barbara Paciorkiewicz spoke quietly and with dignity. But in her story I recognised the truth of our Lebensspuren motto. As Simone Weil had realised, all of us were still desperate to find and connect with our roots. Not being able to – because Lebensborn had destroyed our original identities and because our foster families often erected a wall of secrecy – had eaten away at the fabric of our lives for decades. Barbara spoke for all of us when she said how it had affected her, and how important our new organisation could be.

> All my life I never felt good enough – nor did I know really who I was or where I was truly from. It hurts very much inside. I've always wanted to ask questions but until recently there was no one to ask. Now I want to talk

about this, even though it hurts, so the world does not
forget the terrible idea of stealing children and racial tests.
It must never happen again.

The question was how to achieve this. There were only a few
dozen of us at our first meeting – a fraction of the number of
children born or kidnapped into Lebensborn – and already it
was clear that there were differences between us. Some people
felt we should step out of the shadows and hold a press conference; others wanted to concentrate on building a memorial in
Wernigerode, on the site of the old Lebensborn home. I sensed
then that there would be difficulties in the years ahead.

SIXTEEN | **TAKEN**

*'At 6.30 a.m. about 430 children aged between*
*one and eighteen years were brought by cars to the*
*railway track. The children had only such hand*
*luggage as they could carry themselves. For breakfast*
*they got black coffee and a small piece of bread.'*
GERMAN RED CROSS MEMO, AUGUST 1942

In October 2007, the final pieces of the jigsaw fell into place. The previous two years had been busy. I was still working in my physiotherapy practice (though beginning to think about retiring) and Lebensspuren occupied much of my spare time.

We went public in 2006: around forty Lebensborn children attended our second meeting in Wernigerode, along with journalists from the German and international press. Articles began appearing in quality newspapers and the BBC broadcast a story around the world about what they called *Hitler's Children*. Guntram Weber, Gisela Heidenreich and I answered a never-ending stream of questions, convinced that openness was vital to educate the public about the truth of Himmler's Master Race experiment.

The publicity seemed to work. Gradually it became possible to discuss Lebensborn in public and the more it was talked about, the more enquiries Lebensspuren received from people who suspected they might have been part of the programme. The following year, more than sixty people came to our annual meeting in Wernigerode.

It would be pleasant, though probably naive, to believe it was our frankness that had helped open the archives, which had previously proved so resistant to giving out information. But for whatever reason, formerly unhelpful organisations finally began to open their files. The most important was the International Tracing Service in Bad Arolsen.

The ITS had long been criticised for refusing fully to open up its millions of files. It had claimed, with the backing of the German government, that federal law required one hundred years to pass from the creation of records to their public release. It was an odd argument given that, due to its multinational funding and oversight arrangements, ITS was not technically subject to German law. Its critics alleged that a desire to repress information about the Holocaust in Germany was the real reason for its secrecy.

The fact that in January 2000 all eleven governments sitting on the International Commission of the ITS had endorsed a call for the opening of Nazi archives worldwide seemed to support this view. In March 2006, the US Holocaust Memorial Museum publicly accused the International Tracing Service and the International Committee of the Red Cross of obstruction: 'The ITS and the ICRC have consistently refused to cooperate ... and have kept the archive closed.'

Two months later, the ITS announced that it would finally open its archives in the autumn of 2007.

For seven years I had waited to see the documents it held about me and my family. Two requests had produced only an acknowledgement that there were relevant files but that it needed time 'to evaluate' them, followed by a letter in 2003 asking me for any information I had discovered myself.

When I finally saw them, the documents told me a great deal about both my biological family and Gisela von Oelhafen. The first was a set of papers detailing what had happened to Johann Matko: they showed the dates he had been arrested for partisan activity and held in Mauthausen concentration camp. His name appeared on several lists of political prisoners made by the Nazis: I could not see any reason why the ITS had withheld these from me for so long.

The second tranche of documents came from Lebensborn files: these showed that Ingrid von Oelhafen was once known as Erika Matko from St Sauerbrunn in Yugoslavia. They even included a copy of an insurance policy that Lebensborn had taken out in my name. I had wasted valuable time trying to locate St Sauerbrunn in Austria, while all along the ITS had withheld clear evidence showing I came from Yugoslavia.

The third bundle of records I saw was a series of letters from various branches of the Red Cross dating back to the immediate post-war years. These showed that in 1949 two separate agencies had been searching for me, with a view to returning me to my real family. The first was Caritas, the international Catholic relief organisation, which was responsible for overseeing the work of all welfare organisations accredited to the United Nations. Caritas officials had been tracking children stolen from Nazi-occupied countries – and in particular from Yugoslavia – and must have found my name on the list of transports. They had written to the Yugoslavian Red Cross asking for help in locating me.

At the same time the International branch of the Red Cross, which then had overall responsibility for tracking down and repatriating displaced persons, was also searching for me. It, too, had written to the Yugoslavian Red Cross in an attempt to confirm my true identity.

Surprisingly, given the difficulties caused by the Cold War and their focus on creating a new, unified country, Yugoslavian officials responded. They did not have much in the way of documents themselves – most Nazi files had been destroyed in the fighting or in desperate last-minute bonfires – but they were able to issue an urgent request to the International Tracing Service office in Hamburg. It asked for social workers from the German branch of the Red Cross to visit the von Oelhafen family in Hamburg to ascertain whether I was living with them and whether I was Erika Matko.

I thought back; in 1949 I was not in Hamburg but in the children's home at Langeoog to which Gisela had sent me the day after we escaped to the west. Unless my foster parents chose to help them, the officials would have had no chance of finding and interviewing me. The final document showed that Gisela, at least, was not terribly forthcoming.

GERMAN RED CROSS,
Central Zone, Hamburg

25 October, 1950

To: The International Tracing Service, Child Search Branch, Arolsen

Subject: MATKO, Erika, born on 11/11/1941 in St Sauerbrunn

The wife of von Oelhafen was visited by us. We could not get information from the husband because he does not currently live in Hamburg.

Frau von Oelhafen has been away on a trip for some time. She picked up in person the child identified above from the children's home Kohren-Sahlis at Leipzig and therefore should be able to provide the best information.

However, the only thing she has on paper is a vaccination certificate from Kohren-Sahlis, of which we enclose a copy. Frau von Oelhafen drove to Kohren-Sahlis at the instigation of The Lebensborn Society in Munich. There she was told that Erika Matko was an ethnic German child.

At that point the child Dietmar Holzapfel had already been living with the von Oelhafens for half a year: he had been picked up from the Municipal baby nursing home at Munich. The father of this child is missing at Stalingrad.

Erika MATKO will continue to remain as a foster child with Frau von Oelhafen. An adoption of the child is not intended. Frau von Oelhafen is always happy to make additional information available. She also has a self-interest in learning about the origin of the child, so that she can answer the child's questions. However, she cannot contribute to any further details, having no documents about the child.

The von Oelhafens also fled from the Russians, and in doing so lost their belongings. Frau von Oelhafen remembers that she probably received a transfer note [for Erika Matko] from the children's home at Kohren-Sahlis. This was lost while she was on the run.

We regret that we can currently give you no further information. We would be very grateful if you would

inform us of the outcome of your negotiations in
St Sauerbrunn.

Without Gisela's cooperation – and because Lebensborn had
destroyed all records of my true identity – there was no hope
of finding my biological parents, much less returning me to
them. All the organisations involved abandoned their efforts
at this point.

I was astonished. Gisela had never once told me that the
Red Cross had contacted her – not even during the years when
I was struggling to obtain official documents from the German
government. Nor had she ever even hinted that she knew my
original identity. Worst of all, she had misled the social workers: she had told them that the 'transfer note', given to her and
Hermann when I was handed over to their care by Lebensborn,
had been lost when we escaped from the Russian sector in 1947.
Yet I had found that same document when I cleared out her
room in the 1990s.

I find it hard, even now, to speak badly of Gisela. Deep
down I still want to be loved by her, even though she is long
gone. But I have tried to put aside these emotions, to recognise
this for what it was: a betrayal. She knowingly hid the truth
from me. I was glad – happy, even – to know that someone had
been looking for me all those years ago, but I was very hurt to
learn how that search had been obstructed by the woman I had
once called my mother.

∞

That same month I made my second visit to Slovenia. I organised my trip to coincide with the annual meeting of stolen

children in Celje, and I arranged to travel on from there to Rogaška Slatina, where I was to meet up again with Maria Matko and her family. I was also being filmed for a German television programme: the reporter had arranged for me to spend a day with local historians in Maribor.

I don't know whether it was the presence of the cameras or simply that the Slovenian authorities had managed to unearth more evidence, but my welcome there was warmer than it had been two years earlier, and the meeting much more productive. At last I learned exactly what had happened in Celje – and the truth about Erika Matko.

The occupation of Yugoslavia in the spring of 1941 began well for Germany. Its troops advanced swiftly through the country and in just ten days the Yugoslav High Command capitulated. On 16 April, the day of surrender, the Gestapo moved into Celje and began to arrest local anti-Nazi partisans. Three days later, Himmler himself arrived to inspect the town's ancient prison, Stari Pisker. In the coming months it would be used for the torture and execution of hundreds of resistance fighters.

But unlike other countries overrun by the German Blitzkrieg, Yugoslavia was never fully conquered. Resistance was organised by the charismatic Josip Tito. On 4 July 1941, he issued a secretly printed call to rise up against the Nazi occupiers.

> Peoples of Yugoslavia: Serbs, Croats, Slovenes, Monte-
> negrins, Macedonians and others! Now is the time, the
> hour has struck to rise like one man, in the battle against
> the invaders and hirelings, killers of our peoples. Do not
> falter in the face of any enemy terror. Answer terror with

savage blows at the most vital points of the Fascist occupation bandits. Destroy everything – everything that is
of use to the Fascist invaders. Do not let our railways
carry equipment and other things that serve the Fascist
hordes ... Workers, Peasants, Citizens, and Youth of
Yugoslavia... battle against the Fascist occupation hordes
who are striving to dominate the whole world.

This led to an intensive guerrilla campaign against the Germans.
By September 1941, there were at least 70,000 resistance fighters active in Yugoslavia. Tito's partisans engaged in classic
hit-and-run tactics and when the Germans launched a major
counter-offensive against the rebels, they simply retreated into the
mountains. Hitler's response to this (and partisan activity across
all the occupied countries) was the Nacht und Nebel decree.
The phrase meant 'night and fog': in practice, it was an order to
murder anyone who dared oppose Nazi rule. On 7 December,
the instructions were sent out to commanders in the field.

Within the occupied territories, communistic elements and other circles hostile to Germany have increased
their efforts against the German State and the occupying
powers ...

The amount and the danger of these machinations
oblige us to take severe measures as a deterrent. First of
all the following directives are to be applied:

1.   Within the occupied territories, the adequate punishment for offences committed against the German State or
the occupying power which endanger their security or a
state of readiness is, on principle, the death penalty.

2. The offences listed in paragraph I as a rule are to be dealt with in the occupied countries only if it is probable that sentence of death will be passed upon the offender, or at least the principal offender, and if the trial and the execution can be completed in a very short time. Otherwise the offenders, at least the principal offenders, are to be taken to Germany.

3. Prisoners taken to Germany are subject to military procedure only if particular military interests require this. Should German or foreign authorities enquire about such prisoners, they are to be told that they have been arrested but that the proceedings do not allow any further information.

4. The commanders in the occupied territories and the court authorities within the framework of their jurisdiction are personally responsible for the observance of this decree.

The same day, Himmler issued instructions to his Gestapo and SS forces.

After lengthy consideration, it is the will of the Führer that the measures taken against those who are guilty of offences against the Reich or against the occupation forces in occupied areas should be altered. The Führer is of the opinion that in such cases penal servitude or even a hard labour sentence for life will be regarded as a sign of weakness. An effective and lasting deterrent can be achieved only by the death penalty or by taking measures that will leave the family and the population

uncertain as to the fate of the offender. Deportation to
Germany serves this purpose.

It was a deliberate rejection of the laws of war. No longer would
the Geneva Convention or any other regulations protect the
civilian population of occupied territories. On 12 December,
Feldmarschall Wilhelm Keitel, Chief of the German Supreme
High Command of German armed forces, issued his own re-
enforcement of Hitler's decree to Wehrmacht troops throughout
the expanded German Reich.

> Efficient and enduring intimidation can only be achieved
> either by capital punishment or by measures by which
> the relatives of the criminals do not know the fate of the
> criminal.
>
> In order to nip disorders in the bud the sternest
> measures must be applied at the first sign of insurrec-
> tion. It should also be taken into consideration that in
> the countries in question a human life is often valueless.
> In a reprisal for the life of a German soldier, the general
> rule should be capital punishment for 50–100 communists.
> The manner of execution must have a frightening effect.

But the campaign failed to terrorise the population into
submission. Tito's Partisan Army grew in numbers and effec-
tiveness. By the middle of 1942, Himmler issued orders for
an even greater crackdown on resistance. On 25 June he put
Obergruppenführer Erwin Rösener, Head of SS forces in the
region, in charge of anti-resistance operations and ordered him
to murder or imprison all families suspected of involvement
with the rebels.

Rösener planned six separate *Aktionen*, or campaigns, in the Lower Styria area, centred on Maribor. In the first of these, on 22 July 1942, 1,000 people were arrested and brought to Celje. The men were separated from their families and one hundred of them were lined up against the walls of Stari Pisker, then summarily executed by firing squad. The killings were photographed by Nazi cameramen as a warning against partisan activity. The images were processed at a local photographer's studio: he made extra copies in secret and hid them until the end of the war. It was sobering to hold these pictures in my hands. Among the murdered men that day was Ignaz Matko. Was he one of the bodies lying against the courtyard wall or being lifted on to a stretcher by those who would be shot next?

Some adults were held as hostages, to ensure compliance from the surrounding towns and villages with future *Aktionen*. The remainder were shipped to concentration camps like Auschwitz, where they were murdered or worked and starved to death. Their children were transported to Frohnleiten, in Austria, for racial examinations. Those deemed suitably Aryan – there were not many – were taken away for Germanisation, the rest were sent to 'education camps' where they were treated brutally and suffered from hunger and disease.

The second *Aktion* was scheduled for the beginning of the following month. All families in the nearby villages were ordered to report to the school at Celje on 3 August. Johann and Helena Matko were among the hundreds of families who arrived at the schoolyard that morning with their three children, Tanja, Ludvig and Erika. Heavily armed soldiers quickly separated them into three groups: men, women and children, who were pulled from their parents and taken inside.

Once again, a photographer recorded the event. One picture showed families lined up against the outside wall; another captured the moment when the troops separated them. In this frame a woman in a headscarf was being restrained by a Wehrmacht officer while a second soldier, his rifle slung from his shoulder, stood in front of a mother carrying a baby: she seemed to be pleading with him. A third photograph was shot inside the school: in a crude wooden structure, lined with straw, children and babies were being undressed by unidentified helpers. One little boy was struggling; the faces of others, though, were blank.

Looking at these pictures I felt two quite separate emotions. Somewhere in the crowded yard or among the babies in the schoolroom was Erika Matko: these grainy black and white photographs recorded the day I was stolen from my family. My first reaction was fear. I do not wish to sound melodramatic but I felt a shiver run through my body, a sensation of loneliness and vulnerability. But there was anger, too, at the people who had treated children like this. These soldiers had snatched babies from their mothers and imprisoned them in what amounted to a cattle stall.

Throughout my life I have tried to hide my emotions, to bury them so deeply that the feelings of abandonment and powerlessness could not rise up to engulf me. Those photographs stripped away my defences. I was Erika once again.

The other children and I were kept in the schoolroom for two days. We were crudely assessed for racial value: blond hair, light-coloured skin and blue eyes were taken as a sign of Aryan blood, while brown hair and dark skin or eyes were the marks of worthless Slavonic ancestry. I was designated valuable while Tanja and Ludvig were rejected. They were sent outside and handed back to our parents.

In total, 430 children were judged to be Aryan. We were held in the schoolroom and then taken to the train station. The older children carried babies like me in baskets. The German Red Cross (DRK) was on hand to supervise this children's transport. The DRK's leadership were committed Nazis – even to the point of wearing uniforms and ceremonial daggers – but below them was a civilian army of volunteers. I saw a report by one of these workers, a woman named Anna Rath.

At 6.30 a.m. about 430 children aged between one and eighteen years were brought by cars to the railway track. The children had only such hand luggage as they could carry themselves. For breakfast they got black coffee and a small piece of bread.

The embarkation went smoothly. At 10.30 a.m., a delay of one hour was announced. The train started but only travelled until 2.45 p.m. The children then had to endure a four-hour wait: during this time neither the Red Cross nor officials from the National Socialist Peoples' Welfare Organisation provided any food for them. Fresh water was only brought by a DRK helper who accompanied the train.

Once they arrived in Frohnleiten, the DRK helpers and the young children (two to five years of age) had to walk to the resettlement camp, carrying suitcases and bundles. The children were half-naked and hungry, some with very dirty nappies: this was because there were no clothes to change them. They were screaming and crying.

When we arrived [at the camp], there was another delay for the children because no food was ready for them. They had to stay outside in the courtyard and in a meadow. Finally, at 5 p.m. the children were allowed to go into the eating room. The sixteen DRK accompanying assistants, although tired and weary themselves (they had been on duty since 4.30 a.m.), had to look after the children because all the camp staff, save for four people, were on vacation.

This was how the Nazis treated the children they stole. This was how my life in their hands began. No wonder, then, that I have struggled all my life with a longing to be loved.

In Frohnleiten there were more tests. Himmler's race examiners prodded and poked us, measuring and recording our every characteristic before assigning us to one of four categories. The top two ensured a place in a Lebensborn home; the others guaranteed a ticket to a re-education camp. Among those assessing the kidnapped Yugoslavian children was Inge Viermetz, the female official who had been tried and acquitted at Nuremberg: she was operating on written instructions to 'take only young children who have not yet reached school age'. But it appeared Lebensborn was not the only organisation that wanted us. Officials from VoMI, the agency Himmler had set up to 'protect' ethnic Germans living outside the old Reich, wanted their share of the children who were to become the future leaders of the Master Race. According to testimony given at Nuremberg: 'a real competition for these children arose between VoMI and Lebensborn, and it was thanks to Frau Viermetz's efforts that Lebensborn won in the end'.

I looked at all this evidence and marvelled again at the decision of the judges at Nuremberg to acquit Viermetz and the other senior Lebensborn officials. She had quite literally handled stolen goods: she had decided the fates of hundreds of children, from all over the occupied countries – sending some into a programme that erased their identities and the rest to camps where many died. How could this woman have been declared innocent?

I was moved from Frohnleiten to another camp, at Werdenfels, near Regensburg in Bavaria. At the end of 1942, Lebensborn sent Emilie Edelmann – Gisela Heidenreich's mother – to supervise further racial selection tests. She must have decided that I was good enough: the forms were signed and I was shipped off to Kohren-Sahlis.

Ours was not the last transport of Yugoslavian children. The documents I saw in Maribor revealed that there had been four further *Aktionens* in which hundreds more families were summoned to Celje for racial examination, sifting and separation. They too were divided up between Lebensborn homes and re-education camps.

The officials who showed me these papers had one more revelation for me. Johann and Helena had arrived at the school-yard with three children. When they were permitted to leave, the records showed that they went home with three children – Tanja and Ludvig, and a baby girl called Erika. I knew that my sister and brother had been handed back to my parents, but who was this other child? Somehow Erika Matko had been simultaneously on the train to Frohnleiten and on the journey back to Rogaška Slatina with Johann and Helena. It made no sense.

It was Maria Matko who led me towards the answer. We met in her house the day after my visit to Maribor. After the initial shock of the DNA results she had accepted that I was her sister-in-law. Now, with her help, I was finally able to piece together what had happened to me.

On the day that the children were handed back to Johann and Helena, the Nazis executed several suspected partisans in

Celje prison. Witnesses recalled that the families waiting outside the schoolyard heard the volleys of shots. The children of those murdered men and women were being held with me inside the schoolroom: some were now orphans.

When Tanja and Ludvig were returned to her, Helena must have complained that her third child was missing. Perhaps to appease her, or because they did not know what to do with it, the Germans gave her an orphaned baby. This was the girl who grew up as Erika Matko.

I was torn between pain, anger and bewilderment. My mother must have known that the baby thrust into her arms was not her own. It wouldn't have looked the same nor would it have smelled right, in that indefinable way that mothers know the smell of their own baby. So how could she have accepted this cynical substitution?

The only explanation I could imagine was that she did not dare to argue; that the sound of the firing squads made her fearful for her own life and those of her husband and children. But rational understanding was one thing; emotion quite another. For almost sixty years I had struggled with not knowing who I really was; for the past seven years I had been on a long and harrowing journey to unravel the mystery of my past and discover how I had been transformed from Erika Matko to Ingrid von Oelhafen.

Now I knew. And it didn't help at all.

*'What are we doing? I asked myself.*
*What in God's name are we doing?'*
GITTA SERENY: FORMER UNRRA
CHILD WELFARE OFFICER

I was angry at everyone. Angry with Hitler and Himmler for the orders to kidnap me in the first place; angry with Inge Viermetz and the Lebensborn officials for concealing my true identity and reinventing me as a German child; angry with the soldiers who had given my parents another child in my place. I hated what the Nazis had done to me and to all the other victims of their obsession with pure blood and the Aryan master race. All the resentment and hurt I had suppressed in years gone past was rising to the surface.

My rage was also focused on those much closer to home. Gisela and Hermann von Oelhafen had been willing accomplices in this wretched scheme. They surely should have realised that Lebensborn was not to be trusted: even in wartime Germany there had been enough information – and indeed rumour – about the organisation for them to have had doubts about the provenance of a baby it was offering for fostering.

Then there was Gisela's strange maternal ambivalence: sending me away to foster homes hardly suggested someone who was committed to bringing up a child in a warm and loving environment. Above all, her evasiveness and deceptions about my origins had hampered my search for the truth: how much easier my life would have been if she had only told me where I came from. I knew from my friends in Lebensspuren that some women who fostered Lebensborn children had been open and honest, and that this had eased some of their anxieties. Why did Gisela choose not to talk to me?

But it was the actions of my biological family that hurt the most. I could just about understand why Helena and Johann Matko had accepted the baby handed to them by the Nazis. A family of known partisans could not – especially on a day when their compatriots were being executed – have done anything else. I put myself in their place and tried to imagine the fear of a knock on the door and the discovery of a Gestapo or SS officer standing outside. Although I was haunted by the thought that this other Erika grew up to live what should have been my life, safe in my mother's love, I could not condemn them for it. What I could not accept was that Helena had carried on with the lie long after the end of the war. Barbara Paciorkiewicz's story had shown me that some families of kidnapped children had been determined to bring home their stolen youngsters, yet Helena had lived with the knowledge that her real baby was somewhere in Germany. How could she have carried on without once trying to find me?

I would have loved to ask her myself. But Helena died in 1994: at that time I had not even found the documents Gisela had hidden from me, much less discovered that my roots lay in Slovenia. The actions of my real and my foster mother had conspired to rob me of the chance to seek answers.

My emotions needed a lightning rod: someone living who I could blame for my position. Erika Matko – the other Erika – became the focus of all my anger and pain. Her refusal to meet me, even to answer my letter, enraged me: it seemed so callous. The feeling that she had stolen my life gnawed away at me. Maria had told me that she had been ill much of her life and as a result had never worked. I thought about how hard I had worked to build my physiotherapy business and my struggles with German bureaucracy, and compared this to the way Erika had apparently been supported by her government. And my anger grew.

My friends tried to reason with me. It was not, they rightly pointed out, Erika's fault that she had been given my identity. As a child she could not have known how our lives were swapped and nor could she have done anything about it. And after the war, in Tito's Yugoslavia, surely it would not have been safe for Helena and Johann to reveal any brush with the German occupiers: the communists were not always careful about the innocence or otherwise of those tainted by any form of involvement with the Nazis. Most likely she never knew the truth about her origins; perhaps no one except my parents knew.

Other people urged me to imagine what it must have been like for Erika when I first contacted her. She was then over sixty and suffering from a severe heart condition: it must have been an enormous shock for a complete stranger suddenly to appear and challenge everything she knew about her family. Could I not sympathise?

I could not. I was too consumed by the injustice of everything that had happened to allow myself to feel sorry for a woman whose life I had surely turned upside down.

It took a long while for the anger to dissipate. As months, and then years, passed, I slowly gained enough distance to analyse the situation more clearly. I began to consider what an alternate history would have looked like for me. I thought again of Barbara's story of being taken away from her German foster family and the only home she could really remember; I made myself imagine myself on the train with her to Poland, and feel her bewilderment.

I knew from meeting the group of stolen children in Celje that some of those kidnapped from Yugoslavia had been returned to their families. There had even been a court case to set a precedent for these repatriations. Ivan Petrochik had been snatched by an SS detachment in 1943 when he was less than two years old. His father had been shot by the Gestapo and his mother sent to a concentration camp: he was given the label *Banditenkind* and handed by Lebensborn to a German family.

His mother survived the war and searched for seven years to trace him. In 1952, a court ordered that Ivan be returned to Yugoslavia: he was eleven and had been raised for most of that time as a German child.

Ivan and Barbara's stories made me wonder how the process of repatriation worked and what effect it had on those involved. In 2014, I found some of the answers.

Gitta Sereny was a highly respected journalist and author. She had been born in Vienna in 1921, the daughter of an Austrian aristocrat and a former actress from Hamburg. When she was thirteen, her parents sent her away to boarding school in England, but her train was delayed in Nuremberg and she witnessed one of the Nazis' mass rallies. It left an indelible mark on her and when she finished school she moved to France to help orphans suffering under the German occupation. She also worked with the French Resistance.

When the war ended, she joined the United Nations Relief and Rehabilitation Administration (UNRRA), helping to repatriate the millions of displaced people scattered across the former Reich. She was later assigned to the Child Tracing Department: fifty-three years later she published an account of her experiences in a now-defunct magazine.* In it she described the repatriation of a young boy and girl from Germany to Poland, and for the first time I truly understood what being returned to Yugoslavia would have meant for me.

The process began with Sereny visiting the foster home. It was a traditional Bavarian single-storey farmhouse, its windows uncurtained and with only two dim lights showing the way to the front door.

Sereny had prepared for the visit by examining the region's population records in the local mayor's office. Six people were registered as living at the farm: the farmer and his wife, both in their mid forties, and his elderly parents. There were also two young children – a boy and a girl.

She was keenly aware of the potential distress her visit, and the uncomfortable questions she needed to ask, would cause. It was vital to see the children in the family surroundings, but she hoped that before the interview went too far the youngsters would be sent to bed.

Her reception was distinctly chilly. The family was sat around the kitchen table and deliberately declined to stand up

---

* *Talk* Magazine. Reproduced in the Jewish Virtual Library, 2009. The magazine ceased publication several years ago and Gitta Sereny died in 2012.

when Sereny stepped inside. Although the farmer, his wife and the two children shook her proffered hand – the boy uneasily, the girl enthusiastically – the grandfather refused to do so, hiding his hand behind his back and gruffly demanding what this intruder wanted.

The children were called Johann and Marie. Both were officially six years old and both had blue eyes and blond hair – the boy's cut short and crudely, the girl's longer and neatly braided.

Sereny explained that she only wanted to talk with the family for a short time. To ease the frosty atmosphere she gave each of the children a chocolate bar – a precious gift in the austerity of post-war Germany. It provoked, though, a mixed reaction.

> It was when the little girl, beaming, said, 'Danke', and I stroked her face, that the farmer's wife said sharply, 'Geht zu Bett' [Go to bed], and the two children shot up to obey.

The little girl hugged her mother and reached out for her father's hand. The little boy politely, but formally, bade his parents goodnight then gave Sereny a suspicious look before kissing his grandfather. Then the farmer took the two children away and put them to bed, holding them tightly as he did so.

In 1945 there were 8,500 'unaccompanied children of United Nations and assimilated nationality' registered in the tracing services' files. Within months, tens of thousands of new names were added, sometimes accompanied by snapshot photographs or physical descriptions, all of them kidnapped from the east for Himmler's Germanisation programme. Marie and Johann were among them.

Gitta expressed her disbelief at this situation:

> Who would have taken babies or toddlers away from
> mothers? … How could anyone, even bigots gone mad,
> believe they could discern 'racial values' in young, unde-
> veloped children? Above all how, in practice, could there
> now be large numbers of foreign children – at least some
> of whom would have to be old enough to have mem-
> ories – living, basically in hiding, within the German
> community?

The farmer was hostile when Sereny began asking questions.
He said that their son had been killed by the Red Army during
the siege of Stalingrad; his sister had died four years earlier in a
road accident. They had fostered Johann and Marie to replace
their lost children.

It was plain that this family loved the children and Sereny
tried to reassure the farmers that she understood this, whilst
simultaneously insisting that they must disclose everything they
knew about the youngsters' origins. When she asked about their
biological parents, the farmer's wife said that they had died,
but was very vague about who had given her this information. Sereny pushed harder, explaining to the family that many
Eastern European parents were searching for children who had
been stolen from them.

> 'East?' said the grandfather and, repeating it, virtually spat
> out the hated word: 'East? Our children have nothing to
> do with "east". They are German, German orphans. You
> need only look at them.' And there it was: 'You need only
> look at them.'

Somebody had indeed once looked at them. Just as had happened in Celje, villagers around the city of Łodz had been instructed to bring their children to the Youth Welfare Office where the race examiners had done their work and shipped the chosen children off to Lebensborn. Johann and Marie's parents had been searching for them ever since and had photographs to support their claim. UNRRA decided the children were to be returned to them.

Gitta Sereny was deployed away from the area shortly afterwards. Then, in the summer of 1946, she was sent to work in a Children's Centre in Bavaria. To her surprise – and dismay – she found Johann and Marie were being held there. They were plainly struggling to cope with their removal from the farmer and his wife: both had deep shadows under their eyes and their skin was unhealthily pale. Sereny was shocked by their condition.

> Marie was scrunched up in a chair, her eyes closed, the lids transparent, her thumb in her mouth, but Johann raced up as soon as he saw me, and shouting hoarsely, 'Du! Du! Du!' ['You! You! You!'], hit out at me with feet and fists …

The staff at the centre had seen all of this before: the children's pitiful state was, they told Sereny, all too typical of other youngsters who had been taken away from their German foster families prior to being sent back to their countries of birth. Many, including Johann and Marie, had to be kept in the unit after their official repatriation date; it seemed the only way to ease the pain of what was, after all, the second separation in their young lives and to prepare them for the overpowering expectations of their biological parents. Experience showed that these reunions placed a terrible mental strain on already traumatised children.

This was a caring and thoughtful approach – but for Johann and Marie it had clearly failed. The young boy was already showing signs of aggression, whilst his sister had effectively reverted to babyhood: she wet the bed frequently and would only eat when fed from a bottle.

Later that night, the resident psychiatrist suggested that Sereny try feeding Marie with the bottle.

> She lay there, her eyes shut, the only movement in her lips, which sucked, and in her small throat, which swallowed. I held her until she was asleep. It helped me but, I fear, not her.
>
>   What are we doing? I asked myself. What in God's name are we doing?

Now I understood. This would have been my fate, had I been sent back to Rogaška Slatina. I cannot believe that I would have understood what was happening any more than Johann and Marie grasped why they were taken from the only family they could remember. Now, at last, I wasn't angry any more.

*'My identity might begin with the fact of my
race but it didn't, couldn't, end there.'*
BARACK OBAMA, *DREAMS FROM MY FATHER:
A STORY OF RACE AND INHERITANCE,* 1995

What is identity and how is it formed? Does identity shape the person – or is it the other way round?

This is not, as it might seem, merely an exercise in abstract philosophy. As my journey ended, it was the question I had to face up to. I knew now who I was – or had once been; I was less sure about what this meant.

Identity is much more than merely the answer to the question 'who am I?'. It is also about personality. I was struggling to understand how I had become the person I was today. Was I simply the product of the first years of my life as a Lebensborn child? Was my past to blame for my shyness, my lack of confidence and my desire to put the needs of others – of children especially – above my own? In other words, was the course of my life set in stone by Himmler? That, after all, was what he intended: we Lebensborn babies were supposed to fulfil his vision of a new and uniform generation of the German Master Race.

Surely I was just as much the product of my own choices. Genetics may dictate hair and skin colour, but identity must involve an element of free will. I had chosen to devote my life to working with disabled children; chosen not to get married and have a family of my own. These were my decisions – not the ineluctable result of the Lebensborn programme.

Perhaps those who have never endured the uncertainty of not knowing who they really are, are rarely troubled by these existential questions. And yet which of us hasn't, in our darker moments, returned to a particular moment of our lives and wondered what would have happened had events played out differently?

In Shakespeare's *Hamlet*, Ophelia says, 'We know what we are, but not what we may be.' I could not help but dwell on what might have been. What if I had failed the racial examination that day in Celje? What would my life have been, growing up as Erika Matko? Would I have had the opportunity of a rewarding career, or would my horizons have been limited – as seemed to have been the case with the other Erika – by my environment? If Gisela had been more honest and if the Cold War had not intervened, I would have been reunited with my biological parents: what would that have meant for the course of my life? I asked myself whether I would have been better off had the Nazis left me with my family or whether, in some twisted irony, they did me an eventual favour.

The annual meetings of Lebensborn children exacerbated this uncertainty. The tensions I had detected at our first gathering had grown steadily over the years until Lebensspuren was riven by arguments. All of us had been damaged by our involvement in the Master Race experiment; all of us struggled to come to terms with our personal histories. By 2014, many of those who had joined together to create a supportive environment had walked out or drifted away to form new and smaller groups: I was one of them.

But that year I made two trips which helped me to find some peace. The first was a visit to a former Lebensborn employee, Anneliese Beck. Now ninety-two and almost blind, when I arrived at her home near Frankfurt she greeted me with tea and stollen.

Frau Beck had worked at the Sonnenwiese home in Kohren-Sahlis at the time when I was held within it. She did not remember me: there were 150 youngsters living there, and I had not been in the group for which she was responsible. But she was able to tell me a great deal about the daily routine at Sonnenwiese and to help me understand what my life would have been like. She showed me a photograph of her with some of the children. I was pleased to see that they were nicely dressed and clearly well fed. And she was adamant that despite the circumstances and the presence of the SS, for the most part our time in Kohren-Sahlis was happy and comfortable.

Sitting with Frau Beck helped me fill one of the last remaining gaps in my knowledge. I had no memories of Sonnenwiese, and though I often tried, I was unable to visualise the years I spent there: no matter how much I forced myself to think about it, all I could see was a dark hole. Now that hole was filled and the walls that protected me from my memories were beginning to crumble. I sensed that the final stage would be to travel to Kohren-Sahlis, to walk inside the buildings there: that would, I felt sure, unlock my mind. I was not yet strong enough to go, but I knew that in the coming years I would make that journey.

In October I returned to Slovenia. I went first to Rogaška Slatina where, in a pretty public park, I paid my respects at the memorial to the men and women who were shot between 1941 and 1945. More than one hundred names were carved into the stone: I looked for my Uncle Ignaz and when I found him I traced the letters with my fingertips.

Later, Maria showed me the house where I was born, before taking me to the cemetery, sitting on the top of a hill, where my parents, my grandmother, my brother and my sister are all buried. I laid flowers on the graves and lit candles for my sister and brother, while Maria and her niece cleaned the paths around the stones. I had expected to be overwhelmed by a sense of loss and so I was surprised to discover that, aside from the normal sadness of visiting a graveyard, I felt very little.

It was a similar story later in the afternoon when Maria invited me back to her apartment for Slovenian coffee and homemade blueberry liqueur with other members of the family. The atmosphere was warm, and everyone was hospitable and open. The Matkos had come to accept me as one of their own, and they gave me photographs of my parents, siblings, nephews and nieces. But although I was grateful for the love and generosity of the family I had longed for, I felt like a child among them. My overriding emotion was anxiety, the sort I had always endured when faced with an exam.

The following day I went to the civil registry office to look for records relating to my parents' marriage. The official there dug out a large book in which every local birth was documented. Together we found the page that recorded my arrival; it also revealed that Johann and Helena were married in 1938, several years after my sister Tanja and brother Ludvig were born. It was a clue to the puzzle of the DNA tests: these had shown that while Ludvig's son, Rafael, was definitely my nephew, I was equally certainly not a blood relative of Tanja's son, Marko.

Given that both Tanja and Ludvig had been born before my parents' wedding, the most likely explanation was that Tanja had a different father. The Matko family seemed to have more than its fair share of secrets.

The last remaining mystery was the other Erika. She had still not responded to my letters nor, according to Maria, was she willing to speak to me in person.

I thought long and hard about what I should do. In the end I decided that I would go to her apartment and confront her. I had her address, which turned out to be on the fourth floor of a run-down block in a poor area of Rogaška Slatina. I knew Erika would be at home: the Matkos had told me she was too ill to walk down the stairs and so stayed inside every day. I found her letterbox and the bell with her name on it. I wanted to press it, to be invited upstairs and to see this enigmatic woman with my own eyes. I wanted to hug her, to speak with her and to demand answers. I wanted, above all, to find the peace of mind that would come with confronting my other self.

But I did none of these things. As I stood on the street outside the apartment block, I realised that my emotions were not merely unproductive but corrosive. I had no right to force my own needs on a sick and vulnerable woman who – just as much as me – had been a victim of the Nazis and of Lebensborn. I knew I had to learn not just to understand but to forgive. I walked quietly away.

∞

Two days later, after a last meeting with my Slovenian family, I went home to Osnabrück. As I settled back into my life there, I reflected on what the previous fifteen years had taught me. I seemed to have travelled a great distance, yet my journey had really been a giant circle and I had arrived back at the very place where I began.

It had not been easy or painless, but I was glad – I *am* glad – to know the truth about Lebensborn and how I came to be caught up in it. I draw comfort from the extended 'family' of those who were born or kidnapped into Himmler's experiment; in the years since that first meeting in Hadamar, hundreds of us have discovered what we were searching for.

I am certain that I was once a Yugoslavian child called Erika Matko. I am certain that I was stolen from my family and I am grateful to have had the chance to be reunited with them. I wish, of course, that I could have met my biological mother; I wish I could remember her love for me, and I will always regret not having had the chance to ask her about her life or about why she didn't search for me after the war.

I do not feel close to the Matkos in the way that families should feel close. Too much has happened; too great a separation of time and of place. There is a gulf between us that is more than simply language. I realise that I can no more understand what it means to have grown up in Yugoslavia than I can understand Slovenian. In fact I feel much greater kinship with my late step-brother, Hubertus. We were not blood relations but in our relationship lies the ultimate defeat for the Nazis' ideology: blood is not all-important.

I can smile at that now. How did it take me so long to see something so obvious? I have spent a lifetime working with children burdened with physical or mental disabilities; I have seen how love and patience can overcome these challenges. Nurture can always find a way to beat nature; the hammer does not necessarily shape the hand.

For years I had allowed my life to be overshadowed by the search for something that could not be found. There is for all of us, I believe, a gap between what we want and what we can have, and regret flourishes in that space. I spent too long trapped in a disappointing No Man's Land between dreams and reality. I lost sight of the fundamental truth that we are not defined by the facts of our birth but rather by the choices we make throughout our life.

Mahatma Gandhi once said: 'The best way to find yourself is to lose yourself in the service of others.' It has taken me all my life to understand this. Yet although I am back exactly where I started, if I had not embarked on my journey I don't think I would ever have worked this out. I now know who I was and who I am. Erika Matko was a Lebensborn baby, stolen from Yugoslavia, who disappeared in the madness of the Lebensborn programme. Ingrid von Oelhafen is a German woman, a physical therapist who has brought comfort to generations of children.

My name was Ingrid von Oelhafen. It was also Erika Matko. Ingrid is German; Erika is from Yugoslavia. Both of them were me. But now? Now I am Ingrid Matko-von-Oelhafen. As I always have been.

# AFTERWORD

*'That men do not learn very much
from the lessons of history is the most
important of all the lessons of history.'*
ALDOUS HUXLEY

This is a story of what happened more than seventy years ago. It would be easy to think of it only as history: easy, but wrong.

Since 1945, the world has not known a global conflict. Nor has there been a criminal enterprise on the scale of the Third Reich, nor an ideology that so openly worships the mystic importance of pure blood. But the key words are 'global' and 'openly'. The twisted creed that one person is inherently superior to another by virtue of his race has not disappeared. Nor have the wars fought because of it.

From Southeast Asia to the Middle East, from Africa to the Balkans, there have been those convinced that neighbouring peoples, races or nations are inherently inferior, that these post-Nazi *Untermenschen* are 'other' and therefore less deserving of respect, food, land or life. In the two generations since the Lebensborn experiment died in the rubble of a devastated Europe, the world has known a succession of smaller, more localised conflicts. Most have had, at their root, a version of Himmler's belief in greater and lesser races.

This book is a personal memoir as well as an examination of history. It has been written at a time when the world is fracturing into ever greater hostility between nations, regions or religions. Some of this hostility blossoms into nasty little wars: one ethnic group hacking another to pieces, one branch of a belief system blowing up those it deems to be unworthy in the eyes of its God.

In Europe particularly, and at its borders with countries that were once behind the Iron Curtain, politicians toy with nationalism, stoking the fires of hatred based on racial or historical inferiority. Not since 1945 has the continent – and beyond it, the world – been so dangerously divided.

The lesson of history is that no one learns the lesson of history. It is time we began.

# ACKNOWLEDGEMENTS

Finding my roots has been a long and rocky road, but I have met many wonderful people who accompanied me along the way.

I particularly want to say thank you to my best and most long-standing friend, Dorothee Schlüter. She has been with me from the first timid steps I took in the search for my origins. She has supported me mentally and emotionally and has been deeply involved as I progressed. Thanks are also due to Jutta Schröder, who has always been there with help and care.

I must express my gratitude to Dr Georg Lilienthal for guiding me and to Josef Focks, who pushed me over and over again (as I hesitated) until at last I agreed to go to Slovenia.

To my friends in Lebensspuren, where I first met other children of the Lebensborn programme: you, above all, know how important you have been.

For their company and support on my trips to Rogaška Slatina, I thank my friends Ute Grünwald, Ingrid Rätzmann and Helga Lucas. And I am grateful to my Slovenian family for being so friendly and open.

I owe a special debt to Dr Dorothee Schmitz-Köster. From the moment I met her, she has been a great help. She not only encouraged me to believe that my story could be written but also contributed her extensive knowledge of Lebensborn. She was very good and sensible company on my last trip to Slovenia.

When Tim suggested this book, I made myself examine all the stages of my life. So much had been unknown and troublesome, but as we worked together I found the darkness which had enveloped me gradually disappearing. I also discovered that in the process of writing I was able to 'talk' to Helena, Johann and even Erika Matko. I was able, on the page, if not in reality, to ask 'why'. I did not necessarily find all the answers. But these conversations (some of them heated!) helped me to forgive and to love life as it is.

Ingrid Matko-von-Oelhafen
Osnabrück, April 2015

This book grew out of a film I made in 2013. I had heard of Lebensborn several years earlier and had tried, unsuccessfully, to persuade various television networks to commission a documentary about it. Finally, Channel 5 agreed to fund a sixty-minute film: I am indebted to its commissioning editor, Simon Raikes, for seeing the importance of the story and backing it.

I met Ingrid while researching the programme: she agreed to be filmed and was immensely generous when I was unable, for reasons of space, to include her story within the documentary. She was also kind enough to listen when I subsequently suggested that we should write a book about her extraordinary and brave journey to discover the truth about Lebensborn and her past.

Neither that film nor this book could have emerged without the efforts and encouragement of Dr Dorothee Schmitz-Köster. The Lebensborn children have no greater champion than Dorothee, and her commitment to telling their stories in her own books (sadly published only in Germany) has been crucial in bringing Himmler's shadowy organisation into the light.

Many sources were helpful in cross-checking and verifying the Lebensborn survivors' stories, including: 'Nazi "Selective Breeding"', *The Times* (14 December 1943); 'Hitler's Children', Joshua Hammer, *Newsweek International* (19 March 2000); 'Nazi Past Haunts "Aryan" Children', Kate Bissell, BBC News website (13 June 2005); 'Sixty of Hitler's children meet', Associated Press (5 November 2006); 'Eight people, products of the Lebensborn programme to propagate Ayran traits, met to exchange their stories', Mark Landler, *New York Times* (7 November 2006); 'Nazi program to breed master Race', David Crossland, *Der Spiegel* (7 November 2007); 'Documents detail Nazis' drive for racial purity', Melissa Eddy, Associated Press (6 April 2007); 'Man kidnapped by SS discovers true identity', *Daily Telegraph* (6 January 2009); 'Stolen by the Nazis: The tragic tale of 12,000 blue-eyed blond children taken by the SS to create an Aryan super-race', *Daily Mail* (9 January 2009); 'Stolen Children', Gitta Sereny, *Talk Magazine* (2009); 'Third Reich poster child', Titus Chalk, *ExBerliner* (22 November 2010); Tone Ferenc: Documents relating to the Nazi occupation of Yugolsavia at http://karawankengrenze.at/ferenc/index.php?r=documentshow&id=249

Our thanks are also due to our British publishers, Elliott & Thompson, for so enthusiastically supporting this book, and to our editor there, Olivia Bays: her cool-headed advice significantly improved our manuscript.

Similarly, Andrew Lownie is the very model of a perfect literary agent. His initial guidance, and thereafter his relationship with publishers across the world, has ensured that this story will be read in countries as far apart as Finland, Italy and the United States.

Finally, I could not write without the love and support of my partner, Mia Pennal. After a lifetime of searching, I was lucky enough to be found. *Cursum Perficio*: my journey ends here.

Tim Tate

Wiltshire, April 2015

# INDEX

# ABOUT THE AUTHORS

Ingrid von Oelhafen is a former physical therapist living in Osnabrück, Germany. For more than twenty years she has been investigating her own extraordinary story and that of Lebensborn. She is in contact with other Lebensborn survivors and has been invited to give talks in schools about the programme and its effects on those who were part of it.

Tim Tate is a multi-award-winning documentary filmmaker and author. In 2013 he produced and directed *Lebensborn: Children of the Master Race*, which was broadcast on Channel 5. He is the author of twelve books, including the best-selling *Slave Girl*.